Having Success with NSF

Having Success with NSF

A Practical Guide

Ping Li
Karen Marrongelle

A John Wiley & Sons, Inc., Publication

Cover Image: head © iStockphoto/jpa1999; microscope, DNA, atom © iStockphoto/Kathy Konkle; board © iStockphoto/browndogstudios

Copyright © 2013 by Wiley-Blackwell. All rights reserved.

Published by John Wiley & Sons, Inc., Hoboken, New Jersey.
Published simultaneously in Canada.

No part of this publication may be reproduced, stored in a retrieval system, or transmitted in any form or by any means, electronic, mechanical, photocopying, recording, scanning, or otherwise, except as permitted under Section 107 or 108 of the 1976 United States Copyright Act, without either the prior written permission of the Publisher, or authorization through payment of the appropriate per-copy fee to the Copyright Clearance Center, Inc., 222 Rosewood Drive, Danvers, MA 01923, (978) 750-8400, fax (978) 750-4470, or on the web at www.copyright.com. Requests to the Publisher for permission should be addressed to the Permissions Department, John Wiley & Sons, Inc., 111 River Street, Hoboken, NJ 07030, (201) 748-6011, fax (201) 748-6008, or online at http://www.wiley.com/go/permissions.

Limit of Liability/Disclaimer of Warranty: While the publisher and author have used their best efforts in preparing this book, they make no representations or warranties with respect to the accuracy or completeness of the contents of this book and specifically disclaim any implied warranties of merchantability or fitness for a particular purpose. No warranty may be created or extended by sales representatives or written sales materials. The advice and strategies contained herein may not be suitable for your situation. You should consult with a professional where appropriate. Neither the publisher nor author shall be liable for any loss of profit or any other commercial damages, including but not limited to special, incidental, consequential, or other damages.

For general information on our other products and services or for technical support, please contact our Customer Care Department within the United States at (800) 762-2974, outside the United States at (317) 572-3993 or fax (317) 572-4002.

Wiley also publishes its books in a variety of electronic formats. Some content that appears in print may not be available in electronic formats. For more information about Wiley products, visit our web site at www.wiley.com.

Library of Congress Cataloging-in-Publication Data:

Li, Ping, 1962–
Having success with NSF: a practical guide / Ping Li, Karen Marrongelle.
p. cm.
Includes index.
ISBN 978-1-118-01398-4 (pbk.)
1. Science–Research grants–United States–Handbooks, manuals, etc. 2. National Science Foundation (U.S.)–Handbooks, manuals, etc. I. Marrongelle, Karen. II. Title.
Q180.U5L486 2013
501.4'4–dc23

2012016536

Printed in the United States of America.

10 9 8 7 6 5 4 3 2 1

Contents

Preface

The experience of writing research proposals for funding is essential today for scientists of all ranks. In the process of writing and submitting grant proposals, researchers can better organize their thoughts for exactly what they want to do in a new project, which methods they want to use, how long they plan to do the research, and so on. Even if research funding is not obtained, the exercise of proposal writing is often beneficial to the research process, sometimes almost as important as the actual research itself. Today people are also mindful that there is a huge pressure, especially on early-career investigators, to write proposals and bring grant dollars to the researcher's home institution; one's ability to seek and receive funding from federal grant agencies is often used as a yardstick for important personnel decisions by the institution's administration.

As a major federal funding agency, the National Science Foundation (NSF) has the mission of promoting the progress of science and supporting research and education in the United States. It is a major engine for scientific discovery, learning, and innovation in the United States, and indeed across the globe through its international partnership efforts. NSF is the funding source for about 20% of all federally supported basic research conducted by U.S. colleges and universities, and has played a pivotal role in building the 21st-century research infrastructure for America's science and engineering programs. Scientists across most disciplines can find their home programs at NSF, and as such, it is important for researchers to understand the workings of the NSF review and funding process.

This book grew out of our professional service experiences at NSF. Between 2007 and 2010, the two authors served in their capacities as NSF program directors in the Directorate of Social, Behavioral, and Economic Sciences and the Directorate of Education and Human Resources, respectively. Our NSF services involved the usual duties of a Program Officer, including overseeing program review, managing program budgets, and making informed recommendations of meritorious projects for funding. More important, we had also been actively involved in cross-directorate activities and initiatives, including integrative activities bridging the social/behavioral sciences, education sciences, and computational sciences. Having interacted with many investigators on a daily basis for several years and with many issues, we developed the idea of writing this book, with the hope that the discussion given here will provide a practical guide to investigators, enhancing their understanding of the NSF process and their abilities to write a successful NSF grant proposal.

This book is designed to help researchers, especially early-career investigators, who are interested in applying for NSF grants. Although the specific procedural details may pertain only to NSF, some of the basic principles involving proposal preparation and review may be of general interest to grant proposal writers, no matter where they apply for funding (e.g., we occasionally compare NSF's review process with that of another major federal funding agency, the NIH). This book is not designed, however, to replace researchers' efforts in understanding the specific guidelines laid out by NSF. Researchers who intend to apply for an NSF grant should consult the important information listed on the NSF website (http://www.nsf.gov/) and, specifically, the NSF Grant Proposal Guide. In other words, this book covers issues that relate more to the practical issues in preparing grant proposals, which may not be specified in the guidelines. The book is distinguished from grant writing guides exactly in the "practical" aspect of it, because many of our discussions are based on our personal experiences as NSF Program Officers, from dealing with hundreds of actual questions from investigators or potential investigators and from interactions with colleagues within NSF and across federal agencies. The book includes answers to many questions and provides tips regarding the writing process and beyond, many of which cannot be found elsewhere.

Before our time at NSF, we were both NSF-funded project leaders in research, so we knew about the process of applying for and receiving research funding from the agency. We learned that NSF recruits new Program Officers as "rotators," active scientists from the research community who wish to serve and provide new perspectives to the NSF program management process for an extended period of time, and we were lucky to have the opportunities to serve. We wanted to gain a different type of experience, trying to know what happens on the "other side" of academic research. Clearly, not everyone has the will or the opportunity to spend an extended period of time at NSF as a Program Officer. For those who do have the opportunity to serve, as we did, it means hard work, long periods of absence from family, and significant disruption to active research programs of their own. In writing this book and sharing our NSF experience, we are hoping that it will help the reader to learn how the NSF process works, without he or she needing to have served at NSF. At the same time, we hope that this book may also be helpful to our colleagues currently working at NSF, in that the use of this book, which provides answers to many common questions, could potentially reduce the number of emails or phone calls to those working at NSF.

Given that some changes at NSF may have taken place between the writing of the book (mostly in 2011) and its publication, some of the information presented here may be less accurate or even outdated (e.g., some programs may have been discontinued, while new programs have been initiated). However, we believe that the information provided in the book is mostly up to date, and we have incorporated the newest changes in the revised Proposal and Award Policies and Procedures Guide (PAPPG) of which the Grant Proposal Guide (GPG) is a critical part. The new GPG becomes effective on January 14, 2013 (see http://www.nsf.gov/pubs/2013/nsf13004/nsf13004.jsp), and the most important changes involve the enhanced merit review criteria and guidelines recommended by the National Science Board (NSB; the governing body of the NSF and policy advisors to the President and Congress). We welcome feedback from NSF colleagues, NSF-funded researchers, and anyone who is interested in being an NSF investigator, so that the book could be updated in a timely fashion in the near future.

The reader is reminded here that the opinions, ideas, and views expressed in this book do not represent those of the NSF, as neither the authors nor the publisher have official ties with the NSF. We present the book to the reader as a service to the research community based on our experiences as NSF Program Officers. Readers are strongly encouraged to consult official NSF publications and use this book in ways consistent with its aims and goals. Based on our previous experience and expertise, the division of labor for the writing of the book is as follows: Ping Li wrote Chapters 1, 4, 5, and 7, and Karen Marrongelle wrote Chapters 2, 3, and 6; the two authors have worked together on revising all chapters.

We wish to thank our families for their unfailing support, without which our NSF experience would not have been possible in the first place. Like many NSF rotator colleagues, we had been absentee spouses/parents for many days and weeks during our NSF service, and we thank our families for putting up with this. We would also like to thank our many close colleagues at NSF for their enthusiasm, dedication, and wisdom in jointly running the programs with us in the Directorates of Social, Behavioral, and Economic Sciences (SBE), Education and Human Resources (EHR), Office of Cyberinfrastructure (OCI), and Computer and Information Sciences and Engineering (CISE). To a large extent, this book has been written not by us but by them. Finally, we would like to thank Karen Chambers, Editor of Life Sciences, and Anna Ehler, Editorial Assistant, both at Wiley-Blackwell, who made this project possible and provided kind support throughout the process; Omair Khan, who helped us in formatting the book and providing editorial comments; and Stephanie Sakson and the team at Toppan Best-set Premedia for managing the book composition service. Needless to say, all errors and inaccuracies remain ours.

PING LI
State College, Pennsylvania

KAREN MARRONGELLE
Portland, Oregon

About the Authors

Ping Li is Professor of Psychology, Linguistics, and Information Sciences and Technology, Co-Chair of the Neuroscience Graduate Program, and Co-Director of the Center for Brain, Behavior, and Cognition at Pennsylvania State University. His books include *The Acquisition of Lexical and Grammatical Aspect* (co-authored with Yasuhiro Shirai, 2000, Mouton de Gruyter), *The Handbook of East Asian Psycholinguistics* (Volumes 1–3, General Editor, 2006, Cambridge University Press), *The Expression of Time* (co-edited with W. Klein, 2009, Mouton de Gruyter), and *The Psycholinguistics of Bilingualism* (co-authored with François Grosjean and other guest contributors, 2012, Wiley). He is Editor of the journal *Bilingualism: Language and Cognition*, Associate Editor of *Frontiers in Language Sciences*, and President of the *Society for Computers in Psychology*. He has served as Program Director for the *Cognitive Neuroscience Program* and the *Program in Perception, Action, and Cognition* at the National Science Foundation, as well as Principal Investigator (PI), co-PI, or Consultant for many projects funded by the NSF.

Karen Marrongelle is Assistant Vice Chancellor for Academic Standards and Collaborations at the Oregon University System and Professor in the Fariborz Maseeh Department of Mathematics & Statistics at Portland State University. She has published numerous articles and reports in the area of undergraduate mathematics education research and mathematics professional development. She has served as Program Director in the *Division of Research on Learning in Formal and Informal Settings* at the National Science Foundation, as well as a Principal Investigator or co-PI of many projects funded by the NSF.

Chapter 1

Getting Started

1.1 INTRODUCTION

Scientific research today is quite different from what it was even a few decades ago. Modern science aims at tackling complex problems facing our society, from the understanding of the global environment in which we live, to the origin and development of organisms and life, to the construction of sustainable habitats, and to human behavior and learning. In order to investigate such complex problems, scientists must be engaged in interdisciplinary research, crossing traditional disciplinary boundaries and collaborating with other scientists who have different background knowledge and expertise. Moreover, because complex scientific problems are universally challenging, the approaches and solutions to the problems must also be globally positioned, and as such it requires international perspectives and collaborative engagements. The National Science Foundation (NSF) is a U.S. federal funding agency that promotes scientific research with interdisciplinary and international perspectives.

NSF was established in 1950 by the U.S. Congress with the mission of promoting the progress of science, among other objectives. It supports competitive research projects in all scientific disciplines through a rigorous review process. NSF has seven directorates, including Social, Behavioral and Economic Sciences; Mathematical and Physical Sciences; Geosciences; Engineering; Education and Human Resources; Computer and Information Science and Engineering; and Biological Sciences, which cover the entire spectrum of basic science domains. In addition, it has several offices under the Office of the Director, such as the Office of Cyberinfrastructure and Office of International and Integrative Activities, which also provide funds to support research projects with specific research orientations and tasks (see the NSF Organizational Chart at http://www.nsf.gov/pubs/policydocs/pappguide/nsf13001/ex1.pdf). NSF is the funding source for about 20% of all federally supported research conducted by colleges and universities in the United States. NSF funds reach nearly 2,000 universities and institutions in all 50 states. NSF receives over 45,000 funding requests and makes over 10,000 funding awards each year. In fiscal year 2010, NSF had a budget of about $6.9 billion. Over the last decades, NSF has further differentiated itself from other federal funding agencies, such as the National

Having Success with NSF: A Practical Guide, First Edition. Ping Li and Karen Marrongelle.
© 2013 Wiley-Blackwell. Published 2013 by John Wiley & Sons, Inc.

Institutes of Health (NIH) or the Defense Advanced Research Projects Agency (DARPA), by focusing on supporting basic research at the frontiers of knowledge rather than health- or defense-related projects. This focus, however, does not mean that the NSF is uninterested in research that directly benefits society; on the contrary, NSF is interested in translational and transformative projects that can bridge the gaps between basic research and applied science and technology. In addition to research discoveries, NSF is also keen on supporting projects that can cultivate and educate a new generation of scientists and educators for the 21st century. Toward that goal, NSF provides funds to support science and engineering education, from pre-K, to K–12, to college, and to graduate school and beyond. A major merit index of NSF-funded research is thus the research project's ability to integrate science and education in the research context.

It is quite clear that NSF has been a major engine for scientific discovery, learning, and innovation in the United States, and across the globe through its international partnership efforts (see the latest NSF Strategic Plan at http://www.nsf.gov/news/strategicplan/index.jsp). NSF supported-research has led to the winning of 180 Nobel Prizes, and to many other important scientific discoveries—hence the slogan of NSF: "Where discoveries begin."

This book is designed as a practical guide for researchers interested in applying for NSF funds through its competitive review process. The authors are both active researchers who have received NSF funding and who have both served as Program Officers (sometimes called Program Directors) at the NSF. The overarching goal of this book is to provide investigators with a detailed picture of the NSF application, review, and awarding processes, while at the same time discuss techniques on how to develop and write the best NSF proposals and to maximize chances of funding success. Some of the procedural details may apply only to NSF proposals, but many of the basic principles about proposal preparation and review are applicable to other competitive research proposals. Note that this book is not meant to be a substitute for the official guidelines issued by the NSF. Instead, investigators must closely follow the Grant Proposal Guide (GPG, part of the Proposal and Award Policies and Procedures Guide, last updated on January 14, 2013) in preparing their proposals. Over the course of our discussion, we will cover the basics of NSF proposal preparation, but given the objectives of the book and the space limitation we will not be able to address all the detailed stylistic and procedural requirements.

1.2 FIRST STEPS

Science professionals of all ranks, especially untenured junior faculty in research institutions, are under tremendous pressure these days to seek external funding for their research. This pressure is particularly strong in light of today's business model adopted by many institutions. On the one hand, external funding from a federal or private agency provides the institution of the principal investigator (PI) with increased revenue because of the overhead costs associated with the funding, and on the other hand, it demonstrates to the administrators of the institution (who may be unfamiliar with the PI's specific work) that the PI's research is important because of the rigorous peer review process associated with extramural proposal evaluations. Thus, a researcher's ability to seek and receive external grants is frequently used as an important index for administrators to determine personnel issues such as tenure and promotion. In what follows, we discuss a number of issues you may have regarding the first steps for a grant proposal.

1.2.1 NSF or NIH?

Even before writing a grant proposal, you may be unsure whether you should apply for NSF or NIH funding. Every funding agency, public or private, has its own focuses and priorities, and so do NSF and NIH as the two main federal granting bodies for scientific research. You are strongly encouraged to first study the focuses and priorities of a funding agency before applying. Without a thorough understanding of the funding agency itself, chances of success are slim. Thus, simply knowing that NIH projects are in general funded with more money compared with NSF proposals is not a good reason for applying to NIH (see Section 1.3.4 and Chapter 2, Section 2.4.8, for more on the NSF funding level).

While both NSF and NIH support basic research, their focuses are indeed different. As discussed in Section 1.1, NSF supports basic science, work that elucidates the basic principles and processes underlying the physical world or an organism. NIH, on the other hand, has a mandate to promote public health and welfare and as such puts more emphasis on health-related sciences and applications. In the past, when the funding situation was not so dire, a given proposal often could fit both NSF and NIH goals; today, the division of labor between NSF and NIH has become very clear due to lack of funds, so often you may need to choose one or the other. As an example, a proposal with a significant component on diagnosis and treatment of children's mental disorders would not be appropriate for NSF, whereas a proposal simply on children's developing mental capacities in mathematics or language would not fly at NIH. Of course, this is not to say that you could not have a proposal that would fit the scope and missions of both NIH and NSF (see more discussion below).

Does NSF support work on applications at all? The answer is yes, but the focus is different from that of NIH. In the new merit review guiding principles, NSF further emphasizes projects' potential to benefit society, for which application of basic sciences in achieving societal goals is highly relevant. NSF has a huge interest in integrating science and education, so work that has implications for and applications in the educational setting would be welcome. NSF is particularly interested in supporting work that can boost the education of STEM (science, technology, engineering, and mathematics), as dedicated funds were appropriated by the Congress to support STEM education research. Thus, research on STEM education is a highlighted area for NSF proposals.

What if you feel that your research actually fits both NSF and NIH goals? Can you submit to both funding agencies? Indeed, a proposal that addresses both basic science issues and societal applications may well be entertained by both NIH and

NSF. In such cases, you can simultaneously submit your proposal to both agencies, and it is not illegal or unethical to do so (unlike in the case of submission to peer-reviewed academic journals for publication). It is advisable, however, that you discuss your plan with the Program Officer who is handling your proposal at both NSF and NIH. The Program Officer may give you advice on how to tailor your proposal, even if slightly, to the requirements of his or her agency. On the NSF cover page (see Chapter 3, Section 3.2.2.2) you also need to check the box for "Is this proposal being submitted to another federal agency?" and provide the acronym of the agency to which you are submitting (e.g., NIH, or other federal agencies such as the Department of Education (DOE) or Department of Defense (DOD); see Chapter 7, Section 7.2.2, for further details of these other agencies).

Finally, there may be programs at NSF that are jointly supported by the NIH or other federal agencies to which you can apply. One example is the Collaborative Research in Computational Neuroscience (CRCNS, NSF 11-505), which involves joint NSF and NIH processes (including review and funding efforts). Another example is the Joint DMS/NIGMS (NSF Division of Mathematical Sciences and the National Institute of General Medical Sciences) Initiative to Support Research at the Interface of the Biological and Mathematical Sciences (DMS/NIGMS, NSF 12-561). See Chapter 7, Section 7.2.2, for more on cross-agency collaborations.

1.2.2 Who Can Apply?

Given the importance of extramural funding, a first question that arises is: Who can apply? Different funding agencies tend to use different criteria, some based on eligibility of the investigators, others on research topics, and many based on both. Here we focus on the criteria that NSF uses.

A very important starting point, which has not always been clear to many researchers, is that the official recipient of NSF grants is not the individual PI or researchers, but the institution with which the PI is affiliated. This fact affects the eligibility issue significantly, as it leaves the responsibility for deciding who is eligible for applying for NSF funds in the hands of the institutions rather than NSF. In other words, it is the institution that assigns the PI and that receives the funds from NSF. Indeed, most U.S. institutions have registered with NSF for receiving NSF funding, as well as some foreign institutions such as those from Canada. In the event that your institution is not in NSF's database, this needs to be worked out between your institution and NSF, which is particularly important if you are from a foreign institution (e.g., serving as a collaborator on a project with a U.S.-based PI). Implications of this institution-based grant process also include, for example, that the institution can transfer, terminate, or take other actions with regard to the received NSF funds, although in practice these actions are almost always taken in consultation with the PIs. For instance, the recipient institution, A, can allow the funds to be transferred to another institution, B, in the event that the PI moves to B (see Chapter 6, Section 6.6, for further details on the transfer process). However, institution A can also refuse to transfer these funds to B, and can assign a new PI to the project, if the institution believes that the new PI is perfectly suitable for carrying out the proposed study. In such a situation, the original PI will not have access to the NSF funds proposed in his or her project.

Because of the abovementioned grantor-grantee relationship at an institutional level, it is therefore important for potential applicants to first check with their institution's Sponsored Research Office (SRO) or other units of similar functions before submitting a grant proposal. In general, if you are employed at a four-year college or university in the United States and are a full-time employee at the rank of Assistant Professor or above, you are eligible to apply for funding from NSF. What if you are a postdoctoral fellow associated with a research lab or a research center (sometimes called a research associate)? This becomes a little tricky. First, you should discuss with your postdoctoral mentor to see if he or she is willing to let you apply for NSF funding as an independent PI. Your mentor might prefer to be the PI and have you be the co-PI, so that in case you leave the institution, the funds will remain at your mentor's institution. In some cases (e.g., you move to another institution), your mentor might be willing to transfer some of the funds to you as a subaward (see Chapter 6 for subawards). Second, if you feel strongly about being the PI, make sure that you can carry out the research independently, that is, without your mentor's close supervision and that the work is really yours (of course this needs to be couched tactfully in the face of your mentor). Finally, if you do get an okay from your mentor, you should check with the SRO to see if they are okay with you applying for NSF funding as a Research Associate (a title change may be involved). The institution may want to make sure that by allowing you to be a PI it does not feel obliged to extend your contract.

Another question that often comes up is whether the applicant for NSF funding needs to be a U.S. citizen or permanent resident. The general answer is no, given that the institution with which you are affiliated determines your eligibility, and few U.S. institutions, if any, impose a citizenship or residency requirement on its employees as a criterion for applying for extramural funds. On the other hand, it is important to note that if other personnel such as postdocs or graduate students are part of the proposed project, there may be rules as to whether they need to be U.S. citizens or permanent residents to receive payments. This may especially be an issue with specific programs or solicitations; for example, for the PIRE program (Partnership for International Research and Education), research or travel funds must be used to support only U.S. citizens or permanent residents.

1.2.3 Determining to Which Program to Apply

Once you have determined that you are eligible to apply as an investigator (PI) or co-investigator (co-PI), you should not start the writing of the proposal yet. You should do some more homework before the actual writing. First, understand how NSF works by studying the website http://www.nsf.gov/. This website contains a lot of useful information for potential investigators. For example, it gives an overview of the structure of NSF in terms of the Office of the Director overseeing seven disciplinary Directorates, and Directorates containing Divisions, and Divisions containing Programs. The core unit that evaluates research proposals is at the level of Program, and therefore before you write your proposal you should determine which NSF program or programs you want to apply to, as this will significantly affect how you write your proposal. Even if you have the same ideas, you may be writing quite differently depending on the audience (the Program Officer and panelists). Again, consider the focus and priorities of the specific NSF program to which you apply, just as you do in deciding on the granting agency (NSF vs. NIH; see Section 1.2.1). Often, the match of your research with the profile and priorities of the program may be more important than the actual science itself in getting your work funded.

How do you narrow down your focus onto one or two programs? In practice, this could be quite straightforward. It might be a simple conversation with colleagues in your department, your former advisor, or other experts you know who have overlapping research interests and who have grantsmanship experience. As mentioned in Section 1.1, NSF programs cover a wide spectrum of basic research, and there is a good chance that your research fits squarely with the aims of a specific program. You can determine if this is the case by looking at http://www.nsf.gov/, which contains information about the scope of a program, program contacts, deadlines, and proposal guidelines for different types of applications. It is also important to note here that each program may have a set of program-specific criteria for reviewing proposals, other than the common NSF review criteria (see Chapter 2, Section 2.5).

NSF has a funding opportunity search site ("Find Funding"; http://www.nsf.gov/funding) that is very useful to investigators to identify the opportunities and programs for their applications. You can type in keywords and the website will return with a list of programs that match your search criteria, along with descriptions of the relevant program. You can also search for what has been awarded by keywords, or narrow down your search by specific program areas at NSF. This Find Funding website also provides links to special programs, such as opportunities to support students or postdoctoral fellows, and to recently announced funding opportunities and upcoming due dates.

What if your research project is highly interdisciplinary and does not fit into the scope of one program? As mentioned previously, NSF promotes interdisciplinary and cross-disciplinary research (see http://www.nsf.gov/od/oia/additional_resources/interdisciplinary_research/index.jsp), and, as such, several programs within NSF often work together to co-review and co-fund interdisciplinary projects (see Chapter 4, Section 4.3.3 for more on the co-reviewing process). NSF also often ships out new calls for proposals in response to new demands or trends in science; for example, Cyber-Enabled Discovery and Innovation (CDI), Accelerating Discovery in Science and Engineering through Petascale Simulations and Analysis (PetaApps), Social-Computational Systems (SoCS), and Interface between Computer Science and Economics and Social Sciences (ICES), all of which are jointly supported by multiple units involving computational science, social science, engineering, and education. In some cases, entire divisions have undergone restructuring or realignment to better align the programs with the sciences they represent; for instance, the Division of Chemistry has undertaken a program realignment effort (see http://www.nsf.gov/mps/che/realign/brochure.pdf) in response to changing demands in the field.

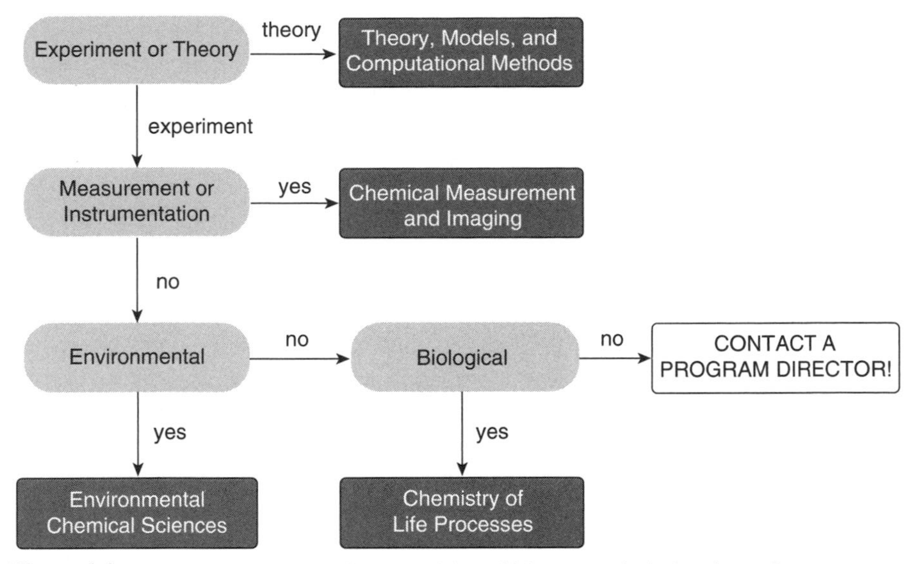

Figure 1.1. Example decision tree for determining which program is the best home for an unsolicited research proposal from a chemistry investigator. Shaded boxes with white lettering are the programs within the Division of Chemistry.

Figure 1.1 provides an example from the Division of Chemistry on how to determine the appropriate home for an unsolicited proposal, a proposal that is written not in response to a specific solicitation or announcement from NSF (although all NSF regular programs do have solicitation numbers).

In sum, it is very important to gather enough information about the program that may be funding your research in the future. If you send your proposal to the wrong program, sometimes a Program Officer may help you transfer your proposal to other programs within NSF (if the Program Officer has the knowledge of where your proposal would best belong), but more often your proposal will end up in the "Decline" category, much like what happens when you send an article to the wrong journal and get rejected right away because it does not fit the scope and aims of the journal. To keep yourself abreast of new developments at NSF, you may want to subscribe to NSF alerts, by clicking on the upper right-hand side of the website http://www.nsf.gov/ (the envelope icon), "Get NSF Updates by Email." NSF will email you information regarding new programs, new solicitations, or Dear Colleague Letters via this alert system (A Dear Colleague Letter [DCL] is often issued to the research community by NSF to tell colleagues that some new changes or opportunities have arisen. See more discussion in Section 1.3.2).

1.3 BEFORE WRITING THE PROPOSAL

In some ways the process of getting a research proposal funded is like the process of getting a paper published, as mentioned above, but in other ways these two processes are different. Notably, getting a research proposal funded involves more homework to be done other than the writing of the proposal itself. In this section, we will discuss a few more steps that are essential, or in some cases just helpful, to the success of a research proposal.

1.3.1 Where to Ask for Help

It's interesting that Figure 1.1 ends with the box "Contact a Program Director" when the decision tree cannot lead you to an appropriate program. Indeed, it is often a good idea to contact the Program Officer even if you have determined the program you want to submit to (see Chapter 2, Section 2.2, for further discussion). Program Officers are in general very helpful and receptive to questions. On the other hand, they would be much happier to answer your questions if you cannot find the relevant answers on the NSF website first, which shows to them that you have done your homework prior to contacting them.

It is probably best to contact the Program Officer first by email and then followed by a phone conversation (they are all busy individuals, often traveling, and may be handling the program while doing some research). You can start your email with something along the following lines (break these into several paragraphs, as appropriate): "*Dear Dr. [insert name of Program Officer], I am in the process of preparing an NSF research proposal in the area of [insert research area]. I am wondering whether I can make an appointment to speak with you by phone to discuss the appropriateness of my proposal for your program. If you prefer to have the discussion by email, that would be fine with me too. I am attaching here a 1-page summary of my research plan. Your comments and suggestions will be most welcome.*" Attaching the summary should be very helpful, as it allows the Program Officer to do a first-pass check on whether your application fits the aims of the program. But remember not to send more materials than necessary—they won't have time to read your whole proposal at this stage, and if they need more information to make a judgment, they will ask you for more (see Chapter 2, Section 2.2, for further discussion).

NSF organizes an annual regional conference as an outreach activity, where NSF Program Officers and staff members provide detailed information regarding NSF and funding. Many investigators who have attended the conference find this meeting useful, as it provides a chance for them to interact with NSF Program Officers and staff in an informal setting (see Chapter 7, Section 7.4, for more details). Other informal settings where you can find NSF staff may include annual meetings of the professional societies in your field; sometimes there are booths or posters from various funding agencies, and you should use these opportunities to make contact with NSF Program Officers. Again, it would be good if you prepare specific questions to ask before meeting with the Program Officers. In some cases, the Program Officer may not have all the information you need but he or she can help you identify relevant information from other programs or sources and establish contacts with other Program Officers at NSF.

Another good resource, as briefly mentioned above, comes from the colleagues in your department, your former mentor or advisor, and other experts you know who have overlapping research interests and who have grantsmanship experience. Some of these individuals may have served on NSF or NIH panels or as Program Officers at federal or private agencies. They can provide you with many useful and insightful tips regarding proposal writing and funding opportunities, at NSF or elsewhere. When there is sufficient research overlap between you and your colleague, it might be a good idea to work on a joint research proposal as a start, although at some point you do want to distinguish your work as an independent line of research and not as part of a larger project by your advisor or your colleague.

A third source of help is your institution's sponsored research office (SRO). Depending on the size of your institution, the SRO may have one or several staff members who are very knowledgeable about NSF and other federal or private foundations. These officers or staff members constantly interact with funding agencies, and some of them may have traveled to funding agencies or to conventions or meetings organized by the agencies (such as the NSF regional conference mentioned above), and have first hand-experience with regard to the process of research funding. They may also be the ones who know about any changes or updates to guidelines, policies, and other requirements (e.g., the new GPG guidelines). Your SRO staff may not have the expertise in your particular domain of research but they have valuable insights in the preparation and funding processes associated with a research grant, and they may also point you to various special opportunities at NSF that you are not aware of.

Finally, you should use the Award Search function at NSF's website (http://www.nsf.gov/awardsearch/) to look for information associated with each awarded project, including abstract, awarded amount, PI names, and the program that provides the funding. In many disciplinary programs' websites, there is a link to "What has been funded," which points to the award search results that pertain to the particular program, providing relevant information on recent awards including abstracts. In some cases, you may want to get in touch with the PI (directly or through NSF) to ask for a copy of the successful research proposal (with personal information redacted). Not every researcher is willing to share his or her proposal in its entirety with the public, but you may be surprised at how helpful some researchers can be in providing valuable information to you. Be aware that NSF Program Officers cannot share copies of full proposals with you; only the PI can do this.

1.3.2 Know the Special Opportunities

Regular programs versus solicitations. NSF has regular programs, as discussed earlier. Each program is associated with a program announcement (e.g., NSF 11-584), available at NSF's website (or you can simply Google the announcement number if you know it). If you apply to the regular program, your proposal is an "unsolicited" proposal. NSF also has temporary and occasional one-time programs, often issued as solicitations. These solicitations are announced to the public via NSF's website and other forms of dissemination (e.g., through institutions' SRO contacts). If you apply to these temporary programs, your proposal is in response to a solicitation, and at the NSF submission website you need to indicate so.

While it is clear from our discussion above that the regular programs may be the main source of funding, you should definitely also pay attention to new solicitations and other new initiatives. Receiving NSF alerts regularly (see Section 1.2.3) may be a good way of being informed about these opportunities, in addition to other means (see Section 1.3.1). For example, NSF recognized the need to bridge the gap between social sciences and computational sciences, and in the last few years it put forward a series of initiatives, such as Cyber-enabled Discovery and Innovation (CDI), Social-Computational Systems (SoCS), and Interface between Computer Science and Economics and Social Sciences (ICES). CDI, SoCS, and ICES were each associated with a solicitation announcement (e.g., 11-502 for CDI, 10-600 for SoCS, 11-584 for ICES), often updated annually (with new solicitation numbers). The CDI program is a particularly interdisciplinary initiative, as it encourages strong collaborations among scientists from computer science, engineering, education, and social sciences. It is important for the potential investigator to read the announcements very carefully to understand the rationale, objectives, eligibility, funding line, timeline, and other requirements associated with each new program or initiative.

One question with respect to the budget of the Program Solicitations is how to read the anticipated funding amount, which is generally specified in the solicitation. The anticipated funding amount is usually the total budget allotted to the new initiative/program for a given academic year (unless otherwise specified), which gives you an idea of how much your own project should aim at. For example, if the anticipated funding amount is $5 million, and the estimated number of awards (usually also specified in the solicitation) is 20–25, then you can put up to $200,000 per year for your project (up to 3 to 5 years, depending on the nature of the program and your project).

Dear Colleague Letters. Another temporary mechanism of funding that NSF uses is the Dear Colleague Letter (DCL). DCLs are also associated with an announcement number (e.g., 11-053 for NSF-DFG collaborative research, and 12-030 *Interdisciplinary Research across the SBE Sciences*), but unlike the regular programs or new solicitations, DCLs are often one-time, having no dedicated new funds or having only limited (supplemental) funds, and in some cases they are simply reminders of changes to existing programs. To give an example, the NSF-DFG Collaborative Research DCL (11-053) announces opportunities for supplementary funds to current NSF awards to enable U.S.-based researchers to collaborate with Germany-based researchers who are funded by the German Research Foundation (DFG), the German counterpart of NSF. The research topics must fit the NSF's Robust Intelligence program and make strong connections to DFG's Autonomous Learning program. To give another example, the HSD (Human and Social Dynamics) program had funds left when the competition was formally completed in 2008, and a DCL was issued to the community that the remaining funds would be used to support large-scale interdisciplinary programs, complexity science projects, and infrastructure projects through existing programs rather than to create new programs in the social and behavioral sciences. Thus, these additional funds augmented the existing programs' capacity to fund more projects in the specified categories, and the projects that received funding might also be larger in size than the regular projects (this is sometimes specifically mentioned in the DCL). In actuality, what happens is that if a proposal receives favorable reviews from an existing program, the proposed project may be supported by funds from both the regular program and the DCL one-time allocations (without the investigators' awareness). Finally, a DCL may simply be a "dear colleague letter," in which an important announcement regarding updates or changes at NSF is made. For example, the most recent and important DCL is 13-004: Issuance of a New NSF Proposal and Award Policies and Procedures Guide (http://www.nsf.gov/pubs/2013/nsf13004/nsf13004.jsp).

1.3.3 Know the Different Types of Support

NSF has a variety of mechanisms to support basic research conducted by different researchers at different career stages. See Chapter 2, Sections 2.8–2.11, for further discussion.

1.3.3.1 Standard Research Proposals

The majority of NSF applications are in the form of proposals for investigator-initiated research projects, regardless of whether you submit to regular programs, targeted solicitations, or other special calls. In the remainder of this book (Chapters 2–5), we will focus on the preparation, submission, and reviewing of this type of proposal.

1.3.3.2 CAREER Awards

NSF used to have small grants for junior faculty members (similar to NIH's academic career awards, or K-awards), but this has now been discontinued. Instead, junior faculty members who are at the untenured assistant professor level can apply for CAREER (Faculty Early Career Development) awards. CAREER awards are research projects, but differ from the regular projects by a strong emphasis on the integration of science and education. For example, you can propose to develop a new curriculum in conjunction with your research. In addition, the CAREER awards are for a duration of 5 years, unlike other regular research projects (usually for 3 years). Although CAREER proposals are reviewed together with regular research projects within disciplinary programs, they are accepted only once a year, typically with a deadline in July (the exact deadline depending on the particular program). See Chapter 2, Section 2.8, for further details on submitting CAREER proposals.

1.3.3.3 Postdoctoral Fellowships

Typically, postdoctoral support is integrated as part of a standard research project, and the investigator has to provide a rationale and a plan for mentoring postdoctoral fellows. In some cases, however, there are special opportunities for postdoctoral support; for example, the Division of Earth Sciences provides generous support to postdoctoral fellows for a period of 2 years, and the recipient of the grant can take the fellowship to an institution of their choice to pursue postdoc research. The Directorate of Biological Sciences has a 1–2 year postdoctoral fellowship in targeted areas of research and teaching in biology, plant genomics, and cross-disciplinary intersections of biology and mathematical and physical sciences. The Directorate of Social, Behavioral and Economic Sciences has a new postdoctoral research fellowship program (SPRF) that includes two tracks: Broadening Participation and Interdisciplinary Research in Behavioral and Social Sciences. The first track was originally the SBE minority postdoctoral research fellowship, whereas the second track is new in 2012 (see details at NSF 12-591). NSF provides a website that contains a list of all postdoctoral fellowship opportunities from various programs (search for "specialized information for postdoctoral fellows" at http://www.nsf.gov/).

1.3.3.4 Graduate Research Fellowships

Like the postdoctoral support, graduate student support is typically built into a standard research grant by the investigator. However, entry-level graduate students can apply, under the sponsorship of their research mentors, to the NSF Graduate Research Fellowship program (GRF). The fellowships will provide 3 years of support to predoctoral students. Only graduate students in their first year or the first term of the second year (the fall semester) are eligible to apply. In addition to the GRFs, NSF also has the Doctoral Dissertation Research Improvement grants in selected programs (see NSF 11-547 for an example). These are usually for 2 years, and unlike the GRFs, these grants are considered supplementary funds (hence smaller awards) and are not full-scale stipends to the students.

1.3.3.5 Conference and Workshop Awards

NSF accepts proposals for conferences or workshops (but not regular conferences of professional societies). The proposal should highlight the uniqueness of the conference or workshop, why it is needed at this time, whether similar conferences have been held before, and a detailed plan of invited speakers, schedule, and dissemination. An important consideration from the NSF Program Officer is whether the proposed speakers or presenters will have chances to get together at regular conferences (not a good thing for the proposal), and if not, whether the proposed conference will lead to interfaces and integration of diverse perspectives and to potential collaborations between the scientists. Unlike the previously discussed proposals, conference proposals do not have to go to panels for review if the total budget does not exceed a certain amount (e.g., $100k), and could be reviewed by the cognizant Program Officers with or without additional ad hoc reviews.

1.3.3.6 Exploratory or Time-Sensitive Grants

There are special smaller-scale grants for exploratory research called EAGER (early concept grants for exploratory research) and RAPID (grants for rapid response research). The EAGER grants are usually for highly novel concepts of research that have not been well established or executed, whereas the RAPID grants are for highly time-sensitive research that studies the effects of a major natural or human event. A good example of RAPID was a proposal for the study of effects that President Obama's taking office in the White House has on the social perception of race and racial stereotyping in the United States. The proposal was submitted to NSF right after the 2008 presidential election. EAGER and RAPID proposals typically do not undergo regular panel review (often reviewed internally by Program Officers only), given their explorative nature and urgency.

1.3.3.7 Other Types of Support

There are many other mechanisms of support at NSF, some of which may be temporary or due to new initiatives. For example, (1) if you have the need to develop infrastructure to acquire a large piece of equipment for research, you can apply for the Major Research Instrumentation (MRI) grant; (2) if you have a large-scale collaborative project involving multiple institutions, you can consider to apply for the Science and Technology Center (STC) grant, which provides multi-million, multi-year support to interdisciplinary and transformative research that can integrate discovery, innovation, and learning at multiple levels, including knowledge and technology transfer; (3) if you have a vibrant interdisciplinary community of science and you are interested in providing new training opportunities to current and future students, you can apply for the Integrative Graduate Education and Research Traineeship (IGERT) grant; (4) if you have strong international collaborations or can forge such collaborations in a given domain and your institution has unique strength in this area, you can apply for the Partnership for International Research and Education (PIRE) grant. Both IGERT and PIRE grants are training grants, which means you will look specifically at how the sciences and the education can be integrated. MRI, STC, IGERT, and PIRE all have restrictions on the number of proposals a given institution can submit to NSF (usually between 1 and 3), so you might be competing with your colleagues internally before you can even send out your real proposal. Finally, NSF sometimes provides supplements to existing grants, typically with smaller funds, and these supplements may be highly program-specific, such as REU (research experiences for undergraduate students) or RET (research experiences for teachers) supplements to projects that have a strong educational component.

Each of the abovementioned types of funding has different requirements with respect to eligibility, goals, research scope, and, of course, budget, and these requirements may also be updated from year to year. It is therefore important for you to learn more about them by reading the NSF website and by contacting the cognizant Program Officers for further information. A good, brief guide by a previous NSF Program Officer can be found in Chapin (2004, Chapter 4: Special Funding Mechanisms), although some of the information there is already outdated.

1.3.4 Budget: How Much to Propose

What is the typical level of funding at NSF? How much money should you ask for in your NSF proposal? These are among the questions most frequently asked by investigators. There are really no standard answers to the budget question, for two reasons: (1) NSF's budget fluctuates from year to year, depending on the economy and on the wills of the White House and Congress, and so the funds each program gets may vary from year to year; and (2) different programs (e.g., each of the types of support discussed under Section 1.3.3) entail different amounts of funding to a project. Thus, it is important to read the program announcements, check with the Program Officers, and look up previously awarded projects (http://www.nsf.gov/awardsearch/) to get a good sense about the possible budget you should put into each proposal. Nevertheless, there are a few simple general rules in guiding your decision on budget.

First, ask for what you need for the aims and the type of your research. In other words, the budget needs to be well justified. If your work can be conducted in a regular lab by you working with a graduate student, do not ask for a special lab technician or for a postdoctoral fellow. On the other hand, if your institution simply does not have the facility or infrastructure support for the type of work you propose, you may want to include in the budget the cost associated with using a piece of equipment at a different institution (e.g., an expensive fMRI scanner your institution does not have). Justification of budget is looked at carefully, especially today when federal funds are so limited. Second, be realistic with your goals of research. Depending on the nature of research you have in mind, you should propose an amount that would allow you to arrive at a reasonable level of research findings. You will probably not solve a big science puzzle about dark matter and dark energy in the universe with $100k per year. Third, make your budget in line with the program's scope and goals, so that you don't propose something that is either too ambitious or too modest for what the program is designed to tackle. This is especially important for the special programs discussed in Section 1.3.3.

In Chapter 2 we will further discuss the eligible costs that you may propose in a proposal, but here we note a few big categories accepted by NSF: personnel (including faculty and student salaries), travel (expenses associated with going to conferences or meetings), equipment (including computers and testing tools), and other direct costs (such as materials and supplies, consultant fees, and publication costs). While many special programs list on the website the total amount each program receives in a fiscal year, regular NSF programs do not list the typical amount of support (not even the cap or range of funding) they would provide to individual projects. Thus, using the above ground rules and these categories of eligible costs, you should consider a standard research grant to be in the range of $300k-$500k for a duration of 3 years. That is, $100k per year is a good ballpark figure to have ($100k as the total cost, direct and indirect together). Given that regular NSF programs typically aim at a 15%-25% funding rate in a good year, the exact amount of each grant will vary considerably, depending on the discipline and the number of proposals received by the program each review cycle (see Chapter 4). In a bad year (e.g., 2010), the funding rate may go down to 10% or lower and proposed budgets are routinely reduced by the program (see more discussion in Chapter 6, Section 6.1).

1.3.5 How Long to Plan for a Proposal

Research grants are hard to come by these days (think about a funding rate of 10% or lower), so the earlier you plan, the better. If you are an entry-level assistant professor, you may be very busy preparing for your courses, getting your lab set up, getting your first students in place, and so on, and may not have time to prepare for a major research grant. But if you can manage, you need to think about extramural finding, if not in your very first year, definitely in your second year. Like most major funding agencies, NSF research proposals undergo a rigorous review process, and this process takes time (more time than most of us would like!). Figure 1.2 gives you a general idea of what's involved in the NSF proposal receipt, review, and funding processes (see also a new dynamic illustration of these processes at the NSF website: http://www.nsf.gov/bfa/dias/policy/merit_review/merit_animation.jsp).

In the remaining chapters we will discuss the details of each of the processes involved, but here it suffices to say that this is a long process. First, after you have spent a long time (at least 90 days) preparing and have submitted your proposal, your proposal finally arrives at the NSF. It then takes about six months for NSF to handle the review of your proposal until the Program Officer recommends your proposal for funding or declination. Then the Division Director has to concur with the Program Officer's recommendation, which also takes time (a few days to 2 weeks, depending on the time of the year). There is an additional period of time after that recommendation for your proposal to go through the administrative hoops (should the recommendation be positive), including processing for the award by the Division of Grants and Agreements (DGA) at NSF. The DGA will usually need about 30 days after the program or division makes a recommendation before it formally notifies the awardee organization (see Chapter 4 for more on these steps). So even if everything goes smoothly for your proposal, you (or more precisely your institution) will have the money in the bank only after some 200 days. More likely than not, your proposal may not sail smoothly through the first time around, and that means that it would be at least a year (or more) between the time you submit your proposal to NSF and the time you receive your grant from NSF (see Chapter 5 for discussion of the revision process). And this kind of timeline is not unique to NSF. In short, the sooner you plan, the better.

In this introductory chapter, we have provided an overview of what NSF is, what the NSF process is like, and some basic steps to do and basic things to know before preparing for an NSF proposal. In Chapter 2, we will discuss the details involved in preparing your proposal; in Chapters 3 through 5, we will discuss the submission, review, and revision processes at NSF; and in Chapters 6 and 7, we will discuss how you can successfully manage your project and leverage NSF funding for further grants. Throughout the discussion we also provide sample letters and various templates for required documents to show with concrete examples the steps that you can follow for having success with NSF. Readers are reminded of the new NSF guidelines effective on January 14, 2013, discussed in Chapters 2 through 6 (see NSF 13-004).

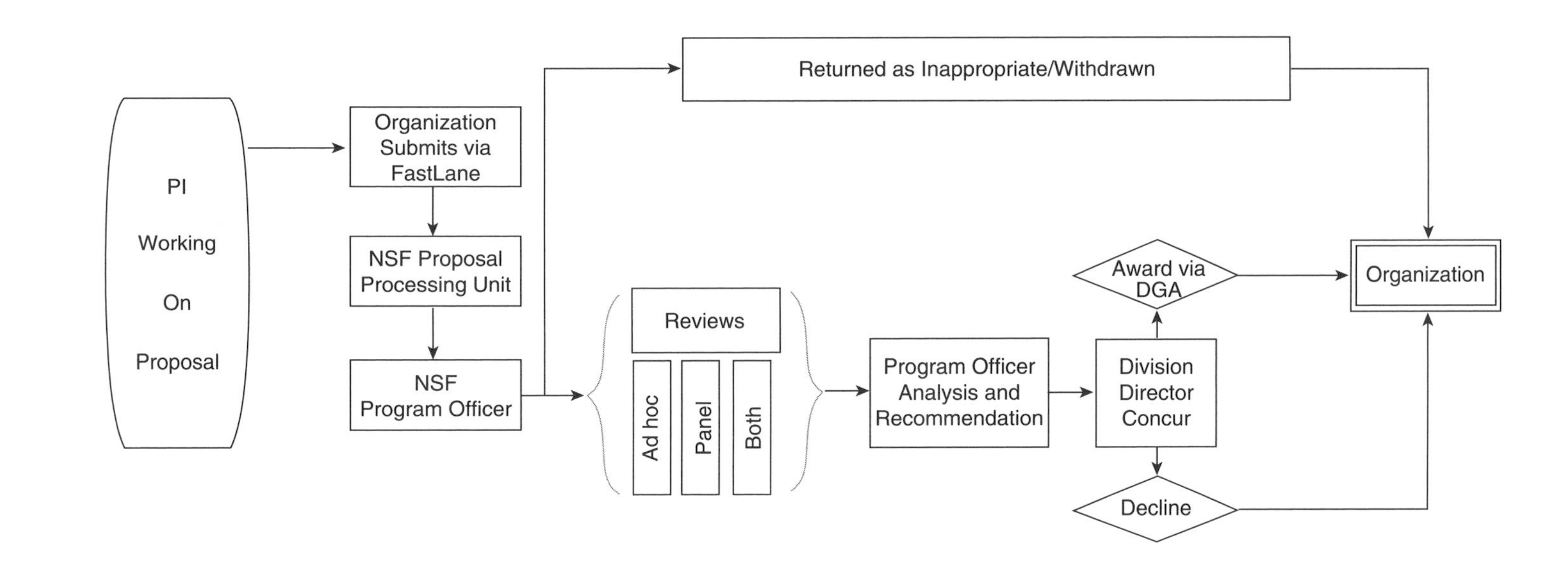

Figure 1.2. NSF proposal process and timeline: (1) The PI and the submitting organization should have at least 90 days for proposal preparation before the target date or deadline. (2) The review and recommendation process usually takes about 6 months. (3) Roughly another 30 days are needed for the paperwork and other processing to complete before the funds reach the PI's organization.

Chapter 2

Preparing Your Proposal

2.1 INTRODUCTION

In this chapter, we will discuss the major components of putting together a proposal for submission to the NSF. Once you have identified a program that fits with your research idea, the next step is to prepare your proposal according to NSF guidelines. One of the best places to start writing is to read and reread the Program Solicitation or announcement (if there is one, as not all programs at NSF have Program Solicitations; see Chapter 1, Section 1.3.2). The Program Solicitation or announcement may contain important information about the types of information to be included in the proposal as well as additional reviewer criteria. You should also continue to work with your Sponsored Research Office (SRO) as you prepare your proposal for submission. You will have questions along the way, some specific to your institution and some more general about NSF. Either way, your SRO will be able to help you find answers to your questions. Generally, you should give yourself or your team about 3–6 months to prepare a proposal for submission to NSF.[1] Competitive proposals are not written overnight and must show that you have "done your homework" by relating to appropriate and relevant literature, understanding ongoing projects funded by NSF, gathering data that you need to make your case, and creating a clearly stated and well-organized document.

We emphasize that before you submit a proposal, you should consult the most recent Grant Proposal Guide (GPG) from NSF. The GPG is NSF's guidebook for the most up-to-date information about what information should be included in your proposal and in what format. The GPG is updated periodically, so you will need to keep your eye out for changes to it (although major changes such as those described in Section 2.5 are rare). NSF sends updates about changes to the GPG through your SRO or via NSF email updates.

[1] Some institutions also require the investigator to notify the SRO office internally (a few weeks to a few months ahead) of the intent to apply for extramural funding, which you should also check.

Having Success with NSF: A Practical Guide, First Edition. Ping Li and Karen Marrongelle.
© 2013 Wiley-Blackwell. Published 2013 by John Wiley & Sons, Inc.

2.2 GETTING TO KNOW THE PROGRAM OFFICERS

It is always a good idea to talk to one or more Program Officers in advance of preparing your proposal for submission (see also Chapter 1, Section 1.3.1). The Program Officers will provide feedback on your idea, help you decide if your idea is a good fit for the program, and be able to answer questions as you prepare your proposal for submission. Some questions you should ask a Program Officer that might impact how you write your proposal include:

- Is my idea appropriate for this program? Are there any other programs at NSF that I should also consider for funding?
- What is the expected composition of the review panel? (Panelist identity is confidential; see Chapter 4, Section 4.2.1.) What expertise has been represented on previous review panels?
- What are the common flaws in proposals to this program?
- Are there specific (or priority) areas that the program is seeking to fund?
- Are there things I should emphasize in my proposal in order to make it stand out?

You will be able to find Program Officers' names and contact information on the NSF website or in the Program Solicitation. Once you have a general idea for your project, write up a "one-pager," a brief, one-page description of your plan. Ask one or more Program Officers to read through your one-pager and provide feedback on the appropriateness of your idea for their program and suggestions they have for improving your proposal. Program Officers like to fund strong proposals. As such, they are often willing to provide guidance and tips on how to write a strong proposal and can provide feedback specific to your brief idea. Program Officers might also connect you with other scholars whose expertise would strengthen and inform your project. Making early, initial contact with a Program Officer is important for a number of reasons: if your idea is not likely to be funded in the program that you have targeted, a Program Officer can tell you this before you have invested time into writing a 15-page full proposal. Alternatively, you might discover in conversations with a Program Officer that your idea is very relevant, timely, and fundable. The Program Officer might then provide tips on writing up your full proposal or suggest other funding opportunities. Finally, in talking with a Program Officer you might discover that your idea is one that involves concepts, target areas, populations, or other components that NSF has funded recently or that the program has received a number of proposals for similar ideas. In a case such as this, the Program Officer might be able to provide guidance on how your proposal can stand out from other, similar proposals.

2.3 HOW TO READ A PROGRAM SOLICITATION: WHERE IS THE IMPORTANT INFORMATION?

The most up-to-date information about proposal preparation is available in the NSF Grant Proposal Guide (the GPG). The GPG is a "living document," so to speak. It is updated regularly (often once per year, although mostly with minor changes). The most recent version of GPG is NSF 13-004 (http://www.nsf.gov/pubs/policydocs/pappguide/nsf13001/gpg_index.jsp), effective January 14, 2013. Your SRO should be familiar with the most up-to-date version, and so should you! The GPG begins with a section outlining significant changes to the Grant Proposal Guide from the previous version and then details all of the information to be included in a proposal to NSF (see the latest GPG Summary of Changes at http://www.nsf.gov/pubs/policydocs/pappguide/nsf13001/gpg_sigchanges.jsp). We must stress here that the discussions in this chapter are not meant to replace your reading and understanding of the GPG, but to provide a starting point and a guide for it. You will use the GPG in concert with a Program Solicitation as you develop your proposal.

2.3.1 Limitations

Program Solicitations will describe any limitations on what types of organizations or researchers can apply for funding under the program. This information is listed under the title of "Organization Limit and PI Limit." Sometimes Program Solicitations restrict the number of proposals that a Principle Investigator (PI) or an organization can submit in a given round of consideration. This information is listed under the "Limit on Number of Proposals per Organization" and "Limit on Number of Proposals per PI" headings. Pay attention to these important sections. For example, if there is a limit of three proposals per organization in a given competition, then you will need to check with the SRO on your campus or in your organization to find out how your institution will decide which proposals will be submitted to the program if there are more than three proposals being developed.[2] Limitations on the number of proposals submitted by a PI might specify how many different proposals you can be involved in (and at what capacity). As an example, in the Frontiers in Earth Systems Dynamics (FESD) solicitation (NSF 10-577), the Program Solicitation states: "An individual may serve as Director (project, center or collaboratory director) on only one FESD proposal (either Type I or Type II), but may be involved in a second proposal in another capacity. No individual may be involved in more than two FESD proposals (either Type I or Type II). The project/center/collaboratory director role is defined in the Program Description section of the FESD solicitation in the description of Type I and Type II proposals."

This statement means that you can submit only one proposal in which you are the director of a project, center, or collaboration. You may also be involved in a second proposal as Senior Personnel (e.g., co-PI, faculty member with time on the project), but you cannot be involved in more than two proposals total.

The first thing to check when you have identified a program or solicitation is what types of submission documents you are requested or required to submit. All

[2] One author's institution forms a special committee (the "downselect committee") each time NSF imposes a limit on the number of proposals allowed for a program. The downselect committee receives project plans from all potential applicants and does an internal evaluation of each project's chances of success according to NSF review criteria. Only the top-rated projects are allowed to be submitted to NSF. In other institutions, this may not be necessary if there is not a huge internal competition.

programs require the submission of a full proposal, but some programs also request or require the submission of additional materials, as discussed below.

Programs at NSF use three types of submission documents: Letters of Intent (LOIs), Preliminary Proposals, and Full Proposals. We describe each of these in turn, and how they differ. LOIs are brief statements (typically one page) that include a list of the PIs, the project title, and a short description of the project. Usually LOIs are used by Program Officers to identify the expertise needed in the reviewer pool before the full proposals are submitted. LOIs are not externally reviewed and you will not receive specific feedback on your LOI from Program Officers. The program announcement or solicitation will indicate whether the LOI is required or optional. LOIs are submitted to NSF via FastLane (see Chapter 3 for discussion of FastLane).

Let's look at an example from the Focused Research Groups in Mathematical Sciences (FRG) solicitation (http://www.nsf.gov/pubs/2006/nsf06580/nsf06580.htm). The Request For Proposals (RFP) provides the following instructions for the preparation of LOIs (Fig. 2.1). Note that for this RFP, a LOI is required. If you think you might submit a full proposal, then you must submit a LOI.

Typically, LOIs are due one or more months prior to the deadline for submission of the full proposal. As you can see from the example above, the information requested in the LOI provides the Program Officers a sense of the type and focus of the project, who is likely to be involved, and which institutions are represented.

2.3.2 Preliminary Proposals

Preliminary proposals are used by NSF programs for two reasons: (1) To strengthen the quality of full proposals and (2) to provide feedback to PIs that might discourage them from submitting a full proposal. Often when a preliminary proposal is used, it is required by the program. That is, in order to submit a full proposal, you must first submit a preliminary proposal and pass the NSF preliminary proposal review process. Preliminary proposals are shorter than full proposals and the specific information to be included in preliminary proposals is outlined in the Program Solicitation. The Program Solicitation will provide very specific instruction about what information to include in a preliminary proposal. Preliminary proposals are submitted to NSF via FastLane, just like the Letters of Intent. Preliminary proposals may be reviewed by outside reviewers in addition to Program Officers. There are two types of feedback given to preliminary proposals: (1) Feedback that is advisory (encourage/discourage) and (2) feedback that is final (invite/not invite). If the feedback is advisory, you will receive a copy of the reviewer's comments, Program Officer comments, and the advisory feedback, either encouraging or discouraging you to submit a full proposal. You may choose to submit a full proposal even if you receive feedback discouraging you to submit! If the feedback is final, you will receive a copy of the reviewer's comments and Program Officer comments, in addition to the decision inviting or not inviting you to submit a full proposal.

The Centers for Chemical Innovation (CCI) Program (http://www.nsf.gov/pubs/2011/nsf11552/nsf11552.htm) provides a helpful example of a preliminary

V. PROPOSAL PREPARATION AND SUBMISSION INSTRUCTIONS

A. Proposal Preparation Instructions

Letters of Intent*(required)*:

Submission of Letters of Intent is required. Each lead proposing organization must submit a Letter of Intent through FastLane. Proposing organizations anticipating the submission of a collaborative proposal should submit only one Letter of Intent from the lead organization.

Letters of Intent will be used by NSF to ensure that the appropriate expertise is available for participation in the review and selection process, to foresee potential conflicts of interest, and to anticipate special award conditions that may be necessary to accommodate the proposed organizational and governance structure. The Letter of Intent is a statement of a proposer's preliminary plans; the senior personnel, collaborating or partnering organizations, and proposed plans may change between submission of the Letter of Intent and submission of the Full Proposal.

Full Proposals may be submitted only by organizations that have submitted a Letter of Intent by the due date, or that have been identified as a non-lead proposing organization in the Letter of Intent for a collaborative proposal.

Letter of Intent Preparation Instructions: Complete submission of a Letter of Intent (LOI) requires two separate components that must each be submitted prior to the LOI due date.

FastLane LOI Component-Via Fastlane, submit the following LOI information:

- Project Title
- Synopsis (a brief abstract of maximum 2,500 characters of plain text)
- Point of Contact for NSF Inquiries
- Project PI Information
- Participating Organizations

Submission of this component via FastLane will produce an **LOI ID** that must be included in the PDF LOI Component described below.

PDF LOI Component-Via an email to the Cognizant Program Officer, submit a document of no more than 5 pages in Portable Document Format (PDF) that addresses the following:

- a description of the proposed organizational structure for the IODP Science Support Office, including the identification of all collaborating and partnering institutions and their roles;
- a list of proposed Key Personnel, including all PIs, Co-PIs and senior personnel, that identifies full names and affiliations;
- a description of the organization's overall management concept for the IODP Science Support Office

Figure 2.1. Excerpt from the International Ocean Discovery Program (IODP) RFP (Request for Proposal) showing instructions for the preparation of Letters of Intent (courtesy of the National Science Foundation).

proposal competition whose results are final. The RFP gives detailed instructions for preparation of a preliminary proposal (see Fig. 2.2).

We highlight a few elements of this excerpt: First, there are specific instructions for the preparation of the Project Description and References Cited for the preliminary proposal that differ from the typical full proposal preparation instructions. In particular, for the preliminary proposal the Project Description is limited to five pages and the References Cited may only include up to 10 citations.

V. PROPOSAL PREPARATION AND SUBMISSION INSTRUCTIONS

A. Proposal Preparation Instructions

Preliminary Proposals *(required)*: Preliminary proposals are required and must be submitted via the NSF FastLane system.

Preliminary proposals (Phase I, required) must be submitted via FastLane by 5:00 p.m. proposer's local time on the due date indicated elsewhere in this solicitation. Preliminary proposals must conform to the format restrictions noted in the NSF GPG and contain only the permitted sections listed below. Note that no supplementary documents are allowed in a CCI Preliminary Proposal.

Cover Page. Please indicate the solicitation number and also check the "preliminary proposal" box. Only the PI's name should appear on the cover page. The budget request should read $1.

Project Summary. In one page describe the intellectual merit and broader impacts of the project. Note that proposals that do not address the intellectual merit and broader impacts of the activity will be returned without review.

Project Description. Limited to 5 pages. CCI preliminary proposals are likely to be read by non-specialists at some stage of the review process. It is therefore particularly important that they be written to emphasize their impact on chemistry in a broad context. The project description should address the following points

- The research challenge to be addressed and the vision for the CCI (both Phase I and Phase II), relevance to sustainable chemistry (approximately 1 page)
- Phase I Research Plan, including the group of initiating investigators, an outline of the initial research goals and how these will link to the Phase II research goals, and other plans to develop the research required for Phase II (approximately 3 pages)
- Brief summaries of plans for innovation, education and professional training, broadening participation, public science outreach, and center management (approximately 1 page).

Reference Section. Up to 10 key references.

Biographical Sketches. Include a two-page biographical sketch for PI and other senior personnel, using the standard GPG guidelines.

Current and Pending Support. List all current and pending research support for PI and other senior personnel.

Single Copy Documents. Single Copy Documents are used by NSF staff, but are not available to the reviewers.

- Suggested Reviewers and Reviewers Not to Include (optional)
- If applicable, a statement excluding other federal agencies from seeing your preliminary proposal and reviews.

Preliminary Proposals will be merit reviewed by ad hoc and/or panel review. The PIs of proposals judged to be meritorious will be invited to submit Phase I full proposals (below).

Figure 2.2. Excerpt from the Centers for Chemical Innovation RFP pertaining to preliminary proposal preparation and review (courtesy of the National Science Foundation).

The RFP states that the results of the preliminary proposal review (invite a full proposal/do not invite a full proposal) are final recommendations. That is, the only way that you can submit a full proposal to the Centers for Chemical Innovation program is to first submit a preliminary proposal and have a successful review in which you are invited to submit a full proposal. Essentially, you must go through two rounds of review to get funded.

You should expect that reviewers of full proposals will be different from reviewers of preliminary proposals. Furthermore, reviewers of full proposals will not have access to your preliminary proposal or reviews thereof.

2.3.3 Full Proposals

The most widely used proposal mechanism at NSF is the full proposal. Full proposals are almost always limited to 15 pages in length. You should approach the writing of your full proposal much like you approach writing a paper for publication in a journal, except that your goal in writing a proposal is to convince a panel of your peers that your project is worthy of funding (see further discussion in Chapter 4, Section 4.1). You will need to address the two NSF review critieria in your proposal: Intellectual Merit and Broader Impacts. You must clearly describe the objectives and significance of your project, describe your methodology and explain why it is appropriate to address the problem(s) undertaken in your proposal. We will discuss Intellectual Merit and Broader Impacts in more detail in the sections that follow (see Section 2.5).

2.4. WHAT MUST I INCLUDE IN MY FULL PROPOSAL?

Your full proposal must include a Project Summary, Project Description, List of References Cited, Biographical Sketch, Statement of Facilities, Equipment, and Other Resources, Statement of Current and Pending Support, Budget, and Budget Justification (all described in more detail below). You will also need to include a Data Management Plan and you may need to include a Postdoctoral Mentoring Plan. The program to which you apply may allow or require supplemental documents, such as letters of collaboration or examples of curriculum.

2.4.1 Formatting Your Proposal

NSF specifies guidelines for acceptable fonts, minimum font sizes, minimum margin spacing, line spacing, and maximum page limitations. Your Program Solicitation or program announcement may have requirements beyond the NSF general requirements. While NSF's only requirement for line spacing is no more than 6 lines of text within one inch of vertical space, most proposals are single-spaced. Keep in mind the reviewers of your proposal as you decide on font size. While NSF allows for 10-point size for fonts such as Arial, Courier New, or Palatino Linotype, reviewers who are reading a stack of 12 or more proposals will not appreciate a small font size (see Chapter 4 for details of reviewers and panelists). In general, you should aim at a clean, well-organized, highly legible format so that your proposal will stand out in style even before the reviewers hit the contents of your proposal.

With very few exceptions, full proposals are limited to 15 pages (including text, tables, and figures). Most competitive proposals use all of the 15 pages and use text that is single-spaced. There is much to write about in a short amount of space; seasoned proposers recognize the need to use all of the space permitted. Typically, you cannot include additional information in appendices (called Supplementary Documentation), but sometimes this is permitted. Be mindful that you don't want reviewers to walk away with the impression that you are trying to squeeze more material than what is allowed into the proposal (see Chapter 5, Section 5.3.3, for more discussion).

2.4.2 Project Summary

The Project Summary is a brief overview of your project and its Intellectual Merit and Broader Impacts. Program Officers and reviewers use the Project Summary sheet to get a sense of what your project is about and what to expect in the 15-page Project Description. NSF has two options for entering your Project Summary information. You may either input your summaries into text boxes in Fastlane (not exceeding a total of 4,700 characters) or, in the case that your Project Summary contains special characters, you may upload your Project Summary as a pdf document. If you enter information directly into the text boxes, you will be prompted to complete three text boxes: (1) Overview of your project; (2) Statement on Intellectual Merit of your project; (3) Statement on Broader Impacts of your project. You must complete all three text boxes or upload a pdf document or FastLane will not accept your proposal. If you upload a pdf document, you must include a section providing an overview of your project, a section describing the Intellectual Merit of your project, and a section describing the Broader Impacts of your project.

A good Project Summary entices the reader to dig into the details of your proposal by presenting specific highlights of the proposed project as they relate to Intellectual Merit and Broader Impacts. A template for your Project Summary is included below:

Overview *Use about one paragraph to give a succinct overview of your project, making sure not to overlap too much with the Intellectual Merit section.*

Intellectual Merit *In this section, you will describe in about a paragraph or two the main thrust of your proposed project and why it is important. Exemplary project summaries describe the science that will be undertaken, why it is important, how it builds from previous work, and why the research team is the right team to conduct the project. This is similar to writing an abstract for a manuscript.*

Broader Impacts *In this section, you will describe in about a paragraph how the project will move the field forward, impact related fields, impact the general public (i.e., societal relevance, if possible), and improve the educational experience of students, including students traditionally underrepresented in the sciences. You can also discuss plans for broader dissemination of your work and for building research or community infrastructure, if applicable. Exemplary Broader Impacts summary statements are realistic, but clearly show the potential for broad contribution. Broader Impacts also relates* *to outreach activities, including public access to science. If you have previously had media exposure of your research (e.g., interviewed by a newspaper, radio, or television), these should be considered broader impact activities. Whether your current work will attract similar media attention will depend on its relevance to society and significance of the work to have broader impacts. See Section 2.5.2 for further discussion.*

2.4.3 Project Description

You will spend the most time developing your Project Description. The Project Description is the 15-page discussion of your ideas, description of your project, and your plans for carrying out your work. The Project Description must include the following items:

- A clear discussion of the work planned
- Objectives of the work planned
- Significance of the work planned
- A separate section with a discussion of the broader impacts of the proposed activity
- Relationship of the proposed work to the PI's/research team's program of research
- Relationship of the proposed work to current states of knowledge
- Preliminary findings obtained, if any
- Results from prior NSF support to the PI or research team, if any.

It is helpful to use headings to organize your 15-page Project Description. You should aim at a consistent heading/subheading system to structure and streamline your discussion. The headings should identify the major components of the proposal. We identify the major proposal components below. (see Chapter 5, Section 5.3.4, for further discussion)

2.4.3.1 Motivation for the Project

You should describe why your project is important for the advancement of science and education, how the project extends the PI's or research team's previous work, and the results of any preliminary studies on the topic. You want to get reviewers and Program Officers excited about your proposal, which involves hooking them on why the project is important and timely and what's new compared to the previous literature (see below). Be mindful, however, that everyone who takes the time to propose a project believes that their project is important and likely to advance science and/or education. You do not want to spend an inordinate amount of time setting the motivation for your project; successful proposals set such rationale in clear yet concise statements. In other words, don't simply say that your proposed work is important (some proposals keep using phrases like "this project is important," "the findings will be important," or "the preliminary data are important"), but show that is important (see Chapter 5, Section 5.3.3, for further discussion). If, after reading your proposal, the reviewer asks "What's the point of this project?" then your proposal is dead in the water even before the panel discussion begins.

2.4.3.2 Relationship of the Project to Relevant Literature

This part of your proposal should show reviewers and Program Officers that you have done your homework. You should situate your project by citing your own background work as well as the work of other scholars that has bearing on your project. You should state how your project would advance or contribute to the current state of knowledge. While this part of the proposal is very important, be careful not to spend too many pages describing background literature. You should provide a synthesis of the most relevant and seminal works, keeping in mind that some reviewers likely will know the literature very well (and will be looking for certain works to be cited), while other reviewers might not know the specific literature as well (so you will want to explain clearly how your proposed project moves the literature forward). Chapter 4 (Sections 4.2.1 and 4.2.2) includes a further discussion of these two types of reviewers and their corresponding roles in the panel and as ad hoc reviewers. Also keep in mind that you can and should cite relevant literature throughout your proposal. Thus, you should think about how to connect important literature to the discussion throughout the Project Description.

2.4.3.3 Description of the General Work Plan and Timeline for the Project Deliverables

You want to spend adequate time in the proposal describing how the proposed work will be carried out, by whom, over what timeline, and any contingency plans you have in case things do not turn out as planned. Many reviewers find it helpful when proposals include a timeline (in tabular format is always a good idea; see Figure 5.2 for an example), outlining and organizing events in the proposed project, including dissemination activities, such as when you expect to write and submit papers for publication and present at professional conferences.

2.4.3.4 Clear Articulation of Methods and Procedures

You should spend the majority of the proposal discussing the methods and procedures for carrying out the proposed work. Such a discussion should include your research questions or research problem statement, the methods you will use for data or sample collection and data/sample analysis, and the expected significance of your results. Many proposals falter because there is not adequate or clear information about procedures and methods to be used, particularly in discussions of data analysis. For example, many educational research proposals give a thorough discussion about how data will be collected, but do not provide specific plans for qualitative data analysis, leaving the reviewers and Program Officers with a vague idea of how the proposed work will be carried out. In your proposal, you must convince reviewers and Program Officers that you are capable of laying out a sound plan for analyzing your data (a plan that is connected to your theoretical framework).

2.4.3.5 Broader Impacts of the Project, Including Plans for Dissemination

NSF requires that you discuss broader impacts of your project (i.e., advancing knowledge, enhancing the infrastructure for research or education, potential benefits for the scientific community and society at large) in a separate section within the 15-page Project Description. It is not expected that your project will address every aspect of the Broader Impacts review criteria; rather, it should address as many aspects as possible but with high quality. At a minimum, the section on Broader Impacts within your Project Description should outline your plans for broadening participation of underrepresented groups in science, and indicating the potential of the proposed activities for societally relevant outcomes. At their essence, Broader Impacts go beyond the scientific impacts of your work (e.g., disseminating your results to your specific research community). True to NSF's mission, Broader Impacts are about the integration of science and education, building infrastructure, dissemination and outreach, the impact of research on society, the involvement of underrepresented groups in science, and being a leader in envisioning the future of science and engineering in line with the NSF Strategic Plan "Empowering the Nation through Discovery and Innovation" (www.nsf.gov/news/strategicplan/nsfstrategicplan_2011_2016.pdf). We discuss Broader Impacts in more detail later in this chapter and in other chapters.

2.4.3.6 Qualification of the PI and Project Team

Provide brief descriptions (about one paragraph each) of the professional/academic background of each senior personnel and why he/she is qualified to conduct the proposed work including the work related to Broader Impacts. Reviewers will use this information to judge the Intellectual Merit and Broader Impacts of the project; in particular, whether or not the PI or team is qualified to carry out the proposed work. Remember that reviewers or panelists may not necessarily be experts in your own field so even if you are a leading figure in your research domain, you still need to have some of this discussion (see Chapter, Sections 4.2.1 and 4.2.2, for information about panels and reviewers).

2.4.3.7 Description of Results from the PI and Co-PIs' Prior NSF Support

No more than 5 pages of the 15-page Project Description should include a discussion of NSF support received in the past 5 years for the PI and all co-PIs, including current NSF-funded projects, irrespective of whether or not salary support was provided. Reviewers will use this information to judge the quality of previous work conducted by the investigators. The GPG (Chapter II: Proposal Preparation Instructions) outlines specific information to be included in discussions of previous NSF support

(NSF award number, title of project, publications resulting from the project, etc.; see GPG section II.C.2(iii)). Summary of Results from Prior Support must address both Intellectual Merit and Broader Impacts in separate sections. You should not only reference publications and presentations, but highlight any outreach or educational activities related to your prior NSF-funded projects. Here the reviewers will evaluate the feasibility of the proposed study in light of what has been achieved. Thus, if you have other preliminary work relevant to the proposed study (but is not funded by NSF), you may also mention it here, for example, by having a subsection on "Other Preliminary Findings."

2.4.3.8 Additional Information

Sometimes Program Solicitations will describe additional components or information to be included in the 15-page Project Description. For instance, in the Discovery Research K-12 program (Division of Research on Learning in Formal and Informal Settings), proposals must include plans for formative and summative evaluation in the 15-page Project Description. The Program Solicitation contains information about formative and summative evaluation and what should be described about the proposed evaluation in the Project Description.

Once you have a draft of your Project Description in hand, it is a good idea to solicit critical feedback from colleagues, mentors, grant support personnel on your campus, and other professionals who may be willing to read and provide suggestions. The more feedback you can receive and incorporate before submitting your proposal, the better! You should aim to submit draft 5+ of your proposal, not draft 1 (see also Chapter 1, Section 1.3.1; Chapter 5, Section 5.3.6).

2.4.4 List of References Cited

You should reference published work, both your own and that of other scholars, after your 15-page Project Description (not included in the 15-page limit). As such, you must include a list of references cited. NSF does not prescribe a particular format for the references cited. You should take care to ensure that all references referred to in your Project Description are included in the Reference list, and the style of citation or reference is consistent throughout. If you have not cited any references in your Project Description, you must provide a statement to that effect and upload it into the "References Cited" section of FastLane. It is highly unusual that you would submit a proposal for review without any cited references; thus, we encourage you to re-think your Project Description if you have not cited any references.

2.4.5 Biographical Sketch

The Grant Proposal Guide outlines specific guidelines for creating a two-page Biographical Sketch. You must follow the NSF format for producing a Biographical Sketch:

(1) Professional Preparation (listed in the following order and format):

Undergraduate Institution(s)	Major	Degree and Year
Graduate Institution(s)	Major	Degree and Year
Postdoctoral Institution(s)	Area	Inclusive Years

(2) List of Academic/Professional Appointments (listed in reverse chronological order)

(3) List of Products Your list of products should be in scientifically acceptable reference format (where applicable and appropriate), with electronic addresses, if available. Acceptable products to list in this section include: publications (articles submitted or accepted for publication are okay to include), patents, copyrights, data sets, software, etc. A good rule of thumb is if your product is citable and accessible, then you may include it. Examples of products that should not be included are presentations and manuscripts not yet submitted for publication. You may include up to 10 products. Up to five of the 10 products should be designated as most closely related to the proposed project. An additional five other significant products can be listed, whether relevant to the proposed project or not.

(4) Synergistic Activities Describe up to five examples of collaborative activity and/or activity that demonstrates past achievements in the area of broader impacts. These could be, for example, the conferences or workshops that you organized or chaired, the editorial services you provided to academic journals (e.g., being an editor, associate editor, or ad hoc reviewer), other professional services (e.g., being a advisory board member, an NSF or NIH panel member, or a student organization adviser or consultant).

(5) List of Collaborators and Other Affiliations NSF Program Officers use information in this section to determine conflicts of interest when choosing reviewers for your proposal. There are three subsections that you must address:

(a) Collaborators and Co-Editors: include in this list any collaborators on a book, article, project, grant proposal, report, or abstract during the past 48 months prior to submission of your proposal or co-editors of a journal, compendium, or conference proceeding during the past 24 months. The list should be alphabetical and include collaborators' institutional affiliations. If you do not have any collaborators that meet these criteria, this should be explicitly stated.

(b) Graduate and Postdoctoral Advisors: Include in this list your own graduate supervisor and any postdoctoral advisors and their current institutional affiliations.

(c) Thesis Advisor and Postgraduate-Scholar Sponsor: Include in this list all students for whom you are/were their major or thesis advisor and any persons, in the last five years, for whom you were postdoctoral sponsor. List current institutional affiliations with all persons.

You must submit a Biographical Sketch for all senior personnel on a project.

2.4.6 List of Facilities, Equipment, and Other Resources

In order to assess resources available to the investigators for carrying out the proposed project, NSF requires a statement of facilities, equipment, and other resources at each research site. You should list only those resources that are directly applicable or available to the proposed project. You will provide an aggregated description of all of the resources, both physical and human, available to your proposed project, including: Laboratory Space, Clinical Space, Access to Animals, Computers, Office Space, Major Instrumentation/Equipment (e.g., fMRI scanner, telescope) and Other Pertinent Space, Equipment, or Resources (e.g., secretarial, electronics shop). Please describe these resources in succinct terms and try not be too lengthy about them in your descriptions. NSF and the reviewers are only interested in whether you have adequate resources to carry out the proposed project, not in how fabulous your facilities or equipment are. If there are no facilities, equipment, or other resources that will be contributed to the project, then a statement to that effect must be included in this section of the proposal and uploaded into FastLane.

2.4.7 Current and Pending Support

NSF requires information about all forms of current and pending support that commits a portion of time for the PI and senior personnel, whether or not salary support is involved. You should list any current or pending grants from NSF, as well as other federal and state agencies (National Institutes of Health, U.S. Department of Energy, State Department of Education, etc.). You can enter your Current and Pending Support information directly into Fastlane. Alternatively, and particularly if you plan to submit a number of grant proposals, you will want to keep an updated document with all of your current and pending projects and upload the document to Fastlane. NSF provides a Word document template for recording Current and Pending Support. You can find this document at www.nsf.gov/pubs/2000/00form1239/00form1239.doc.

For each project that is currently awarded, or pending a recommendation, you must provide the following information:

- Project title
- Whether the project is currently funded, pending funding, or a planned submission
- Source of support (e.g., NSF, NIH, etc.)
- Total award amount
- Duration of the award (start and end dates, either projected or actual)
- Location of the work
- Amount of months per year committed to the project

You will provide this information for all current and pending projects; you must also include the project for which you are applying on the Current and Pending Support list.

2.4.8 Budget and Budget Justification

Next to writing your Project Description, constructing your budget is the other, major component of preparing your proposal. You will want to construct your budget in collaboration with your organization's SRO, as they know the ins and outs of your organization's fiscal policies as well as federal and state policies to which your budget must comply.

The first rule of thumb in constructing budgets is that the amount requested should match the work promised (see also Chapter 1, Section 1.3.4). For instance, requesting $1.5 million to study the practice of six high school science teachers is unlikely to be funded at the full rate. Keep in mind that the budget is your best estimate of what it will cost to do the work proposed. Typically, you can expect that if your proposal enters into negotiations with an NSF Program Officer, you will be asked to trim your project costs. That said, it is not recommended practice to "pad" your budget or overestimate costs. Reviewers and Program Officers carefully consider the amount you have requested for the amount of work promised. During negotiations, you may need to cut down on the work promised if you must cut some of your budget.

You will need to enter a budget for each year of the planned work for your project. For instance, if you plan a 4-year project, you will construct a budget for each of the four years. Every item included in your budget must be justified. So you will prepare not only a numeric budget worksheet, but also a narrative explaining the budget amounts on your worksheet; this narrative is called a Budget Justification and should not be more than three pages long.

It is helpful to construct your budget in a spreadsheet that mimics the NSF budget form. Doing this allows you to play with budget numbers to ensure that you stay within the budget maximum for the program to which you are applying. Items such as indirect costs, fringe benefits, and graduate student tuition remission add up quickly! We describe the components of the NSF budget form below.

A. Senior Personnel Senior Project Personnel are defined as the Principal Investigator, all co-Principal Investigators, and any Faculty Associates, that is, faculty who will participate in the project but are not co-PIs or PI, for example, a Project Manager. Typically, however, Senior Personnel are the PI and co-PIs.

You should list all senior project personnel on the budget sheet, except any senior project personnel who are not taking any person-months or drawing any salary from the project. For each member of Senior Personnel listed, you will list the number of person-months that you expect him or her to devote to the project each year. You may specify either calendar months (typical for those working at institutions on 12-month contracts) or a combination of academic and summer months (typical for those working at institutions of higher education on 9-month contracts). You will need to estimate the amount of effort you think each member of Senior Personnel will devote to the project per month. For instance, if you anticipate spending 20% of your 9-month appointment on the project, then your person-months devoted to the project is 1.8 months (9 months × 0.2). You would then put 1.8 in the Academic Person-Months column on the NSF budget sheet. In order to calculate the amount of funds to request, multiply the number of person-months by your monthly salary. You will do this for each year of the project (your SRO may include a yearly increase of salary by a given percentage adjusted for inflation). Note that your effort on the project may be different in different years.

You may have heard of the "two-month rule": Generally, NSF limits salary compensation for individuals to no more than two months in any one calendar year, across *all* NSF-funded grants. This is different from NIH's rule that allows for up to three summer months or 12 calendar months in the grant support. The reason for this is that NSF regards research as a normal part of the job of faculty members at institutions of higher education and, as such, the faculty member should already be compensated for doing research as part of her/his job. However, there are instances when it is necessary and possible to receive funding from NSF for more than two months in any one calendar year. For instance, if you are not expected to conduct research as a regular part of your faculty workload, and as a result your teaching load does not leave time for conducting research, or if you rely on grant money to pay for your salary (e.g., in a medical school setting), you may cite this in a justification for requesting more than two months of salary in a calendar year.

B. Other Personnel You will need to calculate person-months devoted to the project for postdoctoral associates and other professionals who draw a salary from the project (such as an evaluator or programmer). You will also include the total number of graduate and undergraduate students you expect to support the project and the total dollar amount at which you will support them. Typically, charges for secretarial or clerical work are included in your institution's negotiated indirect cost rate; however, in cases in which projects are sufficiently large, complex, and require an extensive amount of administrative work, you may budget for secretarial and clerical salary. Other personnel who do not fit into the postdoctoral, professional, graduate/undergraduate student, or secretarial/clerical categories are also listed on the "other" line. A description of these "other" personnel must be included in your budget justification.

C. Fringe Benefits Many institutions require that employee fringe benefits—contributions to retirement, health care, social security, and so on—be treated as direct costs. Your institution should have rules and regulations guiding requests for fringe benefits and how your fringe benefits costs are calculated. Typically, fringe benefits are a percentage of the amount of salary requested.

D. Equipment You only need to individually list planned equipment purchases that exceed $5,000 per item. In fact, NSF defines "equipment" as property that costs more than $5,000. Any planned equipment purchase must be properly justified.

E. Travel You may request funds for both domestic and foreign travel in order to conduct fieldwork, consult with colleagues and other experts, and disseminate the results of your work. You should include in a travel budget the cost of air, rail, or car transportation, accommodations, and subsistence. Conference and meeting registration fees may also be included if you are disseminating results of your project at the conference or meeting. You should include in your travel budget the cost of anticipated travel for Senior Personnel, postdoctoral associates, graduate and undergraduate students, and any consultants. Note that travel to Canada and Mexico is considered domestic travel by NSF. If you are requesting funds for international travel, either for fieldwork or for conferences, you should specify the country or countries that you anticipate traveling to, along with further information such as names of conferences, if available. In general, NSF is careful about its money spent outside the United States. You should also be mindful not to request an oversized travel budget, unless the nature of your research requires such a budget.

In general, for a moderate project in the range of $100,000 to $150,000 per year, a request of $2,000 to $3,000 travel budget per year is considered moderate. Amounts higher than this may be considered excessive by reviewers or Program Officers, especially during budget-tight fiscal years.

F. Participant Support Costs If you hold a workshop or provide training and offer a stipend payment for attendees, then you've provided participant support costs. The participant support cost category includes the costs of stipends, transportation, per diem, and other related costs for participants in NSF-supported meetings, conferences, workshops, symposia, and training. You must indicate the number of participants and justify the cost per participant of transportation, stipend, per diem, and other expenses. Normally, Participant Support Costs are not subject to indirect cost calculation; that is, your institution cannot charge indirect costs on any items included in the Participant Support Costs budget lines. However, if you expect to incur significant expenses stemming from the administration of participant support (e.g., to oversee payments to participants), you may negotiate with NSF ahead of time to charge indirect costs on participant support in order to offset expenses associated with participant payment administration. Once Participant Support Costs are awarded, you cannot transfer funds out of this budget line to another budget category without proper justification and the permission of your Program Officer. We discuss proper management of budgets in more detail in Chapter 6 (see Section 6.5).

Note that participant support costs are not the same as payments to human subjects. Payments to human subjects for their participation in a research study should be listed under "other direct costs" (see below).

G. Other Direct Costs Aside from salary, fringe benefits, travel, equipment, and participant support costs, there are often many other resources needed to effectively carry out the work proposed. Other direct costs include: Materials and Supplies (property less than $5,000), costs related to Publication, Dissemination, and Documentation; Consultant Services; Computer Services; Subawards; and other costs that do not fit into any other category (such as payments to human subjects).

Consultant Services at NSF can include fees associated with consultant fees, travel, accommodation, and per diem. Although NSF does not have upper limits on Consultant Service costs, such costs should be estimated reasonably and in accordance with accepted practices and rules used by the researcher's field.

H. Indirect Costs Every organization has an indirect cost rate that is negotiated with NSF or any other federal agency. You must use this negotiated rate to calculate indirect costs charged to the grant. A foreign institution that receives funds from NSF (e.g., as subawardees) is not eligible for indirect cost recovery unless it has a previously negotiated rate agreement with a U.S. federal agency that has a practice of negotiating rates with foreign entities. Indirect cost rates are renegotiated fairly frequently, so it might be the case that you submit a grant proposal with an indirect cost rate of, say, 46%, but once the grant is awarded, your institution's indirect cost rate has been renegotiated at 47.5%. Your institution must honor the original indirect cost rate of 46% for the life of your project.

You or your institution will be responsible for calculating the base of the direct costs from which the indirect costs apply. Generally, indirect costs are calculated on the total direct costs, minus any equipment, participant support costs (for which indirect costs are not allowable, except in unusually costly circumstances; see Section 2.4.8, Part F), and all subaward costs in excess of $25,000. This constitutes the base (sometimes called the "modified total direct cost" by some institutions) upon which the indirect costs are calculated.

We cannot state enough that it is to your advantage to work with someone at your institution (e.g., an SRO staff member) who knows the local, state, and federal guidelines as you prepare your budget. Your own institution might have policies that are more restrictive than the federal guidelines. More important, seasoned budget preparers will be able to help you make the most of your budget request.

A final word on budget: NSF does not have the manpower or resources to oversee the budget operation of most regular research projects once the funds are released to the institution. Thus, it leaves budget issues to the grantee institutions, and it is your and your institution's responsibility to monitor the budget and carry out the research in good faith. NSF has very specific rules laid out in the General Grant Conditions regarding when prior approval is required for change of budget or change of research scope (see Chapter 6, Section 6.5, for more on this topic).

2.4.9 Data Management Plan

All proposals submitted to NSF must include a Data Management Plan, that is, a plan for managing data throughout the project and disseminating findings of your research. Your Data Management Plan will be uploaded as a supplementary document of not more than two pages. NSF has prepared more specific guidelines for certain Directorates, Offices, and Divisions, which can be found at http://www.nsf.gov/bfa/dias/policy/dmp.jsp. Your Data Management Plan should include a description of the types of data, samples, or products that will be produced in your project; any standards that will be applied to the format of the data, samples, or products produced in your project; policies to which you will adhere for access and sharing of data, samples, and products (e.g., human subjects); policies to which you will adhere for reuse of your data, samples, or products; and plans for storing and archiving data, samples, and products and ensuring access to them.

Some investigators may be concerned about the privacy of their data or the need to work on the data for publication before sharing the data with others. It is okay from NSF's perspective if the PI says that the data will be made public after some time has elapsed. NSF does not demand that data be shared immediately with the community after they are gathered. But you do need to discuss ways in which data are stored and ways in which data sharing will be facilitated. In other words, NSF requires that findings and outcome from your project be accessible to benefit the broader scientific community rather than just your own lab.

A sample Data Management Plan template follows.

Data Management Plan

The proposed project will adhere to NSF's policy with regard to data sharing and management. The following steps are taken to ensure efficient and effective data storage, sharing, protection, and management.

1. Products of the Research and the Types of Data. [Describe the data the project will produce and how the data will be generated/collected; for example, do you have data from children, adults, and the elderly, do you have metadata that include experimental protocols, demographic data, experimental stimuli, etc.]

2. Data Format Descriptions. [Describe how the data will be coded, entered into databases, transformed (e.g., transcribed), and de-identified if applicable.]

3. Data Sharing Practices and Policies. [Describe how the data will be accessible to the scientific community and to the public, and any practices or policies governing data sharing including the use of web-based tools for online access; if you need to have the data accessible to the community only after the publication of the relevant study, say so here.]

4. Data Reuse, Redistribution, and Production of Derivatives. [Describe opportunities, if any, for reanalysis and reuse of data; issues of participant privacy, intellectual property rights, and other liability issues in connection with the reuse and redistribution of data should be discussed here; also specify if the data are provided only to the scientific community and not for commercial development.]

5. Archiving of Data. [Describe how the data will be archived, stored, backed up, and protected.]

2.4.10 Postdoctoral Researcher Mentoring Plan

In one page or less, you must describe your plans for mentoring any postdoctoral researchers supported by the project. In this plan you should specify details with regard to how you will mentor the postdoctoral fellow, the expectations of the postdoctoral fellow, and any plans for structured feedback to the postdoctoral fellow. Examples of mentoring activities might include (but are certainly not limited to):

- Opportunities for independent research
- Training in the preparation of presentations, publications, and grant proposals
- Opportunities for supervising graduate or undergraduate students
- Career guidance
- Training in responsible and ethical conduct of science and professional practices
- Opportunities to collaborate with researchers from diverse backgrounds and disciplinary areas
- Opportunities for positive career trajectories and professional development.

Some universities provide resources for mentoring graduate students and postdoctoral fellows. If your institution provides such resources, you should consult with them as you develop your Postdoctoral Researcher Mentoring Plan. You must only submit this document if you request support for postdoctoral researchers. Like the Data Management Plan, your Postdoctoral Mentoring Plan will be uploaded as a supplementary document.

A sample Postdoctoral Researcher Mentoring Plan template is below:

Postdoctoral Researcher Mentoring Plan

We request funds to support a [insert full- or part-time] postdoctoral researcher to participate in the [insert appropriate parts of project] for [insert duration of project].

The postdoctoral researcher will have expertise in [insert description of experience and skills]. He or she will play a significant role in [insert project activities].

The proposed project and the academic environment in which the project is carried out will provide significant opportunities and positive mentoring experiences to the postdoctoral researcher. [Describe mentoring experiences here.] In particular, the postdoctoral researcher will work closely with the PI in [describe activities here, such as formulating theoretical hypotheses, developing research paradigms, presenting at professional meetings, preparing grant proposals, and preparing manuscripts for publication]. The postdoctoral researcher will also be given opportunities to supervise graduate and undergraduate students, and to teach or co-teach with the PI in [insert course titles] to increase his or her teaching experience. Finally, the postdoctoral researcher will have ample opportunities to interact with other researchers who share similar research interests with the PI, including colleagues from [insert research domains or department/college names]. [Insert other activities in which the postdoctoral researcher will engage and how these activities will benefit his/her career trajectory.] The goal is to have the postdoctoral researcher leave the program as a productive scholar with a track record ready for independent research and teaching at the tenure-track assistant professor level in a college or research institution setting.

2.4.11 Letters of Commitment

When you invite other researchers to participate in your project, in an advisory capacity or consulting role, you will want to include letters outlining their commitment to participate. (Note that few NSF programs allow for letters recommending funding for a project, that is, a letter stating that the project has merit and is worthy of funding.)

Commitment letters should be written on official letterhead and addressed to the project PI. It is standard practice to send a template of the commitment letter to those who have agreed to advise or consult on your project, asking the letter-writer to add in customized content where appropriate. We provide a sample commitment letter below.

January 3, 2012
Dear Dr. Principal Investigator:

I wish to offer my strong support for your proposed project, [Insert Project Title]. As a [insert discipline specialization researcher/scholar/faculty], I recognize the need for [the specific activities of the project].

This project offers the opportunity to [describe how the field of study will be enhanced and expanded]; your research is timely and important [insert reasons why]. Few research studies [do what the proposed study plans to do]; the proposed study is well-designed and an appropriate step for understanding [insert specifics].

I believe my experience and expertise will be valuable to your project. [list here the qualifications and previous activities that are pertinent to the proposed project and why this person is qualified to be an advisor/consultant on the project.]

I am pleased to accept your invitation to serve on the project's [advisory board, consultant group, etc.], recognizing that the total time commitment is approximately [x number] days per year in a mix of face-to-face and virtual meetings. I understand that my role in this project will involve the duties of [list specific tasks to be performed]. I will look forward to collaborating with you on this important project.

I wish you good luck with your NSF application and every success with your research.

Yours truly,
Esteemed Professor

It is preferable to submit commitment letters on official letterhead and with a signature. However, it may be acceptable to submit an email correspondence in cases when a colleague cannot provide a formal letter (e.g., the person is in a remote location for an extended period of time).

2.5 INTELLECTUAL MERIT AND BROADER IMPACTS

All proposals to NSF are evaluated according to two merit review criteria approved by the National Science Board (NSB; http://www.nsf.gov/nsb/): Intellectual Merit and Broader Impacts. NSF reviews all proposals in terms of three guiding review principles, two review criteria, and five review elements (see below). If you previously submitted proposals to or reviewed for NSF, these review principles and elements may be new to you. However, the core criteria that guide the evaluation of

NSF proposals remain as Intellectual Merit and Broader Impacts. Reviewers are asked to structure their reviews according to these two criteria. Program Officers structure their comments according to these two criteria. The evaluation of all proposals to NSF, regardless of division or program, is grounded in these two merit review criteria. All NSF solicitations contain statements of these two criteria and five review elements for proposers and reviewers. The following five review elements are considered in the review for both Intellectual Merit and Broader Impacts:

1 What is the potential for the proposed activity to:
   a. advance knowledge and understanding within its own field or across different fields (Intellectual Merit); and
   b. benefit society or advance desired societal outcomes (Broader Impacts)?
2. To what extent do the proposed activities suggest and explore creative, original, or potentially transformative concepts?
3. Is the plan for carrying out the proposed activities well-reasoned, well-organized, and based on a sound rationale? Does the plan incorporate a mechanism to assess success?
4. How well qualified is the individual, team, or institution to conduct the proposed activities?
5. Are there adequate resources available to the PI (either at the home institution or through collaborations) to carry out the proposed activities?

Furthermore, three Merit Review Principles should be given due diligence by proposers, as reviewers will consider these three principles when reviewing your proposal:

1. All NSF projects should be of the highest quality and have the potential to advance, if not transform, the frontiers of knowledge.
2. NSF projects, in the aggregate, should contribute more broadly to achieving societal goals.
3. Meaningful assessment and evaluation of NSF-funded projects should be based on appropriate metrics, keeping in mind the likely correlation between the effect of broader impacts and the resources provided to implement projects.

2.5.1 Addressing Intellectual Merit in your Proposal

Intellectual Merit is a phrase generally referring to the scientific significance, rigor, and creative activities contained within a proposal. When you write up your 15-page Project Description, you should take care to address (1) how the project will advance knowledge, (2) your or your team's qualifications to carry out the proposed project, (3) that you have access to the resources (people, supplies, equipment, space, etc.) required to carry out the proposed project, and (4) the extent to which your project explores ideas that are original, creative, or potentially transformative. Reviewers and Program Officers recognize that a proposal that is well written and well organized usually reflects a project that is well conceived and well thought out. NSF reviewers are asked specifically to consider *what* the proposers want to do, *why* they want to do it, *how* they plan to do it, *how* they know if they will succeed, and *what* benefits would accrue if the project is successful. These questions will be applied to both the technical aspects of the proposal and the way in which the project may make broader contributions.

NSF has increasingly focused on the "transformative potential" of a proposal. Transformative potential is reflected in research activities that suggest or explore creative, original, or potentially transformative concepts, ideas, or research paradigms. Such activities could revolutionize entire disciplines, create new fields, or disrupt accepted theories and perspectives. You need to "think outside the box" in your research projects to demonstrate innovative thinking that could lead to paradigm-changing ideas. Such projects may appear to be risky at first but may yield high pay-offs. They may also involve unique collaborative or partnership efforts that would create synergies across disciplines and lead to new discoveries. Incremental research, as opposed to transformative work, may be important in its own right, but given NSF's limited budget these days, incremental research has low priority of funding as compared to transformative projects.

2.5.2 Addressing Broader Impacts in Your Proposal

Broader Impacts refer to the potential of a project to affect changes in the scientific community, educational arenas, and society more generally. A central part of NSF's vision is to advance education by "cultivating a world-class, broadly inclusive science and engineering workforce, and expand the scientific literacy of all citizens" (NSF, 2006, p. 5)[3]. As such, Broader Impacts should go beyond disseminating information to scientific communities, but also include K-20 educational components and groups typically underrepresented in science and engineering, and should develop messages for society at large. NSF believes that integration of research and education and broadening participation of underrepresented groups are core strategies in line with NSF's Strategic Plan, hence critical components of Broader Impacts. In the newly revised merit review criteria, NSF further stresses the importance for funded projects, in the aggregate, to contribute broadly to achieving societal goals, that is, to go beyond the intrinsic values of advancing scientific knowledge (see further discussion below).

Broader Impacts tend to be the more challenging component of a proposal to develop and communicate. In a recent review of the application and interpretation of the merit review criteria by PIs, reviewers, and Program Officers, NSF found that the Broader Impacts review criterion is not generally well understood and that the criterion has not been consistently implemented by reviewers (see http://www.

[3] http://www.nsf.gov/pubs/2006/nsf0648/NSF-06-48_4.pdf.

nsf.gov/nsb/publications/2011/meritreviewcriteria.pdf). Some scholars (e.g., MacFadden, 2009) argue that because scientists receive little training in developing and carrying out broader impacts activities, this merit review criteria is both poorly understood and only weakly implemented. Broader Impacts means more than simply disseminating the results of your research to your academic field of study, although dissemination of results is a very important part of your work plan. Broader Impacts involve, for instance, the integration of science and education, building infrastructure, dissemination and outreach, the impact of research on society, and the involvement of underrepresented groups in science.

NSF has provided guidance about and examples of Broader Impacts at http://www.nsf.gov/bfa/dias/policy/merit_review/resources.jsp and in the Grant Proposal Guide. We encourage you to refer to this important document as you develop your plans for addressing Broader Impacts. In general, your plan for addressing Broader Impacts should be well organized, well thought out, and well integrated with the Intellectual Merit of your project. You should not include a "laundry list" of disconnected activities in your Broader Impacts work plan. For instance, Broader Impacts might include inviting underrepresented minority high school students into your lab to assist with your research or giving a public-friendly talk about your research at a "Science Pub." You might consider partnering with other faculty or community members to help develop your Broader Impacts work plan. For example, you might partner with a staff member at a local science museum to develop a plan to get your work into the community in an understandable way. Some national organizations provide support and guidance for researchers who want to disseminate their research to the public. For instance, the American Association for the Advancement of Science (AAAS) offers resources for researchers to communicate more effectively with the public (http://communicatingscience.aaas.org/). Think creatively about partnering with others to develop and implement your Broader Impacts work plan. Note that these examples are for illustrative purposes; instead of providing a prescriptive list of requirements, NSF now leaves it to the investigator to include concrete examples of Broader Impacts relevant to the aims and scope of the proposed project.

As of January 2013, NSF requires a separate section in every Project Description to describe the Broader Impacts of the proposed project. In this section, you should include clearly stated goals for your Broader Impacts activities, a description of the activities you plan to do, and your plan for documenting the output of those activities. Your plan for Broader Impacts should be not only creative but well organized and based on a sound rationale. Furthermore, you or your research team should be qualified to carry out the activities described in your Broader Impacts work plan and you should have sufficient access to resources necessary to carry out your Broader Impacts work plan. To reiterate, Broader Impacts involve specifically the abilities of the proposed project in activities that benefit society, such as the integration of science and education, building of infrastructure for research and education, broad dissemination and outreach, the impact and benefit of research to society, and the involvement of underrepresented groups in science. Some further dimensions to consider Broader Impacts, according to a report from the National Science Board (see http://www.nsf.gov/nsb/publications/2011/meritreviewcriteria.pdf), could include (but are not limited to) increased participation of women, persons with disabilities, and underrepresented minorities in science, technology, engineering, and mathematics (STEM); improved STEM education at all levels; increased public scientific literacy and public engagement with science and technology; improved well-being of individuals in society; development of a globally competitive STEM workforce; increased partnerships between academia, industry, and others; increased national security; increased economic competitiveness of the United States; and enhanced infrastructure for research and education. These examples, just as other examples of Broader Impacts discussed above, however, are not meant to be exhaustive or prescriptive, and should not be thought of as a "check list" to be included in every NSF proposal. Investigators may include other appropriate societal outcomes not covered by these examples.

2.5.3 Are Intellectual Merit and Broader Impacts Weighted Equally in the Review Process?

Both Intellectual Merit and Broader Impacts are given full consideration during the review process. Each criterion is necessary, but neither is sufficient. As such, you should try to make the strength of each of these criteria similar in your proposal. Competitive proposals spend the majority of their pages communicating the nuts and bolts of the proposed project—such descriptions often contain elements of intellectual merit and broader impacts. You need both strong intellectual merit and broader impacts to be funded at NSF, but without a strong scientific project (Intellectual Merit), your chances of being funded are greatly reduced. That is to say that Broader Impacts alone will not carry your proposal. Likewise, if your proposal has very strong Intellectual Merit but weak Broader Impacts, your reviewer ratings are likely to reflect the weak Broader Impacts, again reducing your chances of being funded. We will say more about the two merit review criteria of Intellectual Merit and Broader Impacts in Chapter 4 when we discuss the review process (see Chapter 4, Section 4.1).

2.6 SUPPLEMENTARY DOCUMENTS

Unless a Program Solicitation specifically asks for or allows supplementary documents (such as letters of collaboration and examples of curriculum), you should not include such documents with your proposal; if you do, your proposal will be returned without review. In addition, try not to include URLs to websites at which you have posted additional information relevant to the project, which will be viewed as another way of squeezing more material into the proposal than allowed. Panelists and reviewers are unlikely to go to individual websites to look for additional information, given the load of reviews that they typically take (see Chapter 4).

2.7 COLLABORATIVE PROPOSALS

Much work funded by NSF is collaborative. Often, collaborations happen across organizations and in these cases, the investigators must decide upon the mechanism by which to submit a proposal to NSF. There are two main ways to submit a proposal for collaborative work to NSF. First, you and your co-investigator(s) can submit a single proposal for your work. One institution would be the lead and submit the proposal; the proposal would contain subawards to the other institution or institutions. The second option is for you and your co-investigators to simultaneously submit proposals from your separate institutions that contain the same documentation except for the budget and budget justification. That is, the Title, Project Summary, Project Description, and References Cited are the same for all of the proposals. Information unique to each institution—that is, budget and budget justification, biographical sketches, current and pending support, and Facilities, Equipment, and Other Resources—will be submitted separately by all institutions involved. The major difference between these two options comes in the process of submitting the proposal and then executing the project once it is funded.

Let's consider the two options with an example: Professor Smith at the University of Oregon, Professor Jiang at Boise State University, and Professor Walters at University of Dayton are planning to submit a proposal to NSF together. Professor Smith wants to submit a single proposal from the University of Oregon with subawards to Boise State University (for Professor Jiang's work) and the University of Dayton (for Professor Walters's work). Professor Smith will be listed as the PI of the project, and Professors Jiang and Walters will be listed as co-PIs. Professors Smith, Jiang, and Walters all work together to write the Project Summary and Project Description, and generate the References Cited and develop the budget collaboratively. Professor Smith will work with her SRO to submit the grant from the University of Oregon. This means that Professor Smith and her SRO will upload all of the grant documents: Project Summary, Project Description, and References Cited. Professor Smith will collect Biographical Sketches and Current and Pending Support from Professors Jiang and Walters and submit them, too. Professor Smith will also be responsible for completing the Facilities, Equipment, and Other Resources page to reflect the resources at all three institutions. Professor Smith also will upload any Supplementary Documentation (e.g., Data Management Plan). Professor Smith will upload the project budget from the University of Oregon, and Professors Jiang and Walters will upload their subaward budgets (which will need to be approved by their universities) directly. We will describe in more detail the process of submitting your proposal electronically in Chapter 3.

Alternatively, Professor Smith, Professor Jiang, and Professor Walters each submit a proposal from their own institutions. They are each listed as PI on their proposals. On the project cover page, they must start the title with "Collaborative Research:" followed by the actual title of the project. The project needs a lead proposal, and Professor Jiang has agreed to do this; as such, Professor Jiang and Boise State University upload the Project Summary, Project Description, and References Cited (because the proposals will be linked online, Professors Smith and Walters's proposals will have the same Project Summary, Project Description, and References Cited, even though they do not individually have to upload them). Professor Jiang will also be responsible for uploading any Supplementary Documents. Each of the PIs will upload their own Biographical Sketches, Current and Pending Support, Facilities, Equipment, and Other Resources, and their own Budgets and Budget Justifications.

As you can see from the above examples, in the first case, where a single proposal was submitted for the group's work, Professor Smith and the University of Oregon assume the bulk of the work uploading the proposal (the NSF proposal will be a single integrated file in one piece). The University of Oregon will have to coordinate with Boise State University and the University of Dayton to execute the subawards or subcontracts. The budget, essentially, is controlled at just one institution: the University of Oregon.

In the second case, each institution acts more or less independently, uploading their own documents and budgets. These are essentially three individual proposals, linked online as one Collaborative Research project (see Chapter 5, Section 5.3.5, for involving colleagues in a Collaborative Project). In addition, Professor Jiang (or one of the other two PIs upon agreement) will upload the Project Summary, Project Description, References Cited, and any Supplementary Documents for the set of proposals. When the project is funded, each institution will receive its own award from NSF and be responsible for managing budgets and reporting.

There are pros and cons to submitting a single proposal or a set of linked, individual proposals, which will be described in more detail in Chapters 3 and 5 (e.g., Chapter 5, Section 5.3.5).

Next, we describe details about applying for some specialized funding opportunities at NSF that were mentioned in Chapter 1, Section 1.3.

2.8 FACULTY EARLY CAREER DEVELOPMENT PROGRAM

As described in Chapter 1, the Faculty Early Career Development Program (CAREER) is open to tenure-track faculty (or equivalent) who have not yet been recommended for tenure. The CAREER program operates NSF-wide; that is, all Directorates at NSF participate in the CAREER program, so whether you are a theoretical physicist or a marine biologist, you are eligible to apply. Note that different NSF divisions can support different numbers of CAREER grants. Smaller programs, such as the programs in the Social, Behavioral, and Economic Sciences Directorate, typically fund very few CAREER proposals each year (because the programs in this Directorate can be small and the budgets for CAREER grants can be quite large in comparison). By contrast, a Division such as the Division of Information and Intelligent Systems funds a number of CAREER proposals each year (10 or more). Generally, you will prepare your CAREER full proposal like you would prepare a proposal to any other program except for the following important differences:

- You will need to include an education plan in addition to a research plan and a plan to integrate research and education. The so-called Integrated Research and Education Plan can be challenging to develop and communicate. You might consult NSF's advice on developing an excellent plan for Broader Impacts (see Section 2.5.2) as a starting place. Note that in addition to Intellectual Merit and Broader Impacts, reviewers will be instructed to rate your proposal on the Integrated Research and Education Plan.
- You cannot have co-PIs or other senior personnel on your CAREER proposal. You may include other researchers in your project, but only as consultants or advisory board members.
- You must include a letter of support from your department that includes acknowledgment that you are eligible for the CAREER program, the department's support of your research and development, and an acknowledgment of the merit of your project. The CAREER solicitation outlines specific requirements for the departmental letter.

NSF provides a thorough Frequently Asked Question document about the CAREER program at http://www.nsf.gov/pubs/2011/nsf11038/nsf11038.jsp#b21. We encourage you to carefully read this document if you are planning on applying for a CAREER award.

2.9 RESEARCH EXPERIENCES FOR UNDERGRADUATES

NSF supports the inclusion of undergraduate students in research projects and, as such, has a solicitation specifically targeted to this result. There are two tracks for Research Experiences for Undergraduates (REU) awards: (1) a stand-alone project that involves the development of an REU site to engage undergraduate students in a variety of research projects and (2) an REU supplement to an existing NSF-funded research project designed to engage undergraduate students in research. Applying for an REU Site requires the development of a full, 15-page proposal outlining the objectives of the REU Site program, how undergraduate students will be trained in conducting research, the types of research projects, and how students will be recruited and selected, particularly students from underrepresented groups. You will want to follow the general guidance in preparing a full 15-page proposal, but also consult carefully with the REU Solicitation, as the solicitation contains important instructions for proposal preparation. REU Site proposals have a specific deadline (typically over the summer) and are accepted once per year.

If you have an existing NSF grant, you may request an REU Supplement to your existing award (or as part of a new or renewal proposal). As with any supplement to an existing award, you will want to talk with your Program Officer before submitting an REU supplement request (see Chapter 7, Section 7.2.1, for more on submitting supplements). You will describe the justification for an REU supplement in three pages. Your justification should include a description of the activities the students will undertake in the research project, your experience involving undergraduate students in research, and how students have been or will be selected. Note that REU supplements typically can support up to two students, but exceptions may be made for involving a larger number of students from underrepresented groups or large research efforts. The REU solicitation provides some budget guidance, but it is always a good idea to consult with your Program Officer to see if your budget is "in the ballpark." If you read the REU solicitation carefully, you will notice that you can also request an REU Supplement as part of a proposal for a new or renewal grant. The REU solicitation provides specific instructions on how to embed a request for an REU Supplement into your full proposal. Basically, you will add a Supplementary Document to your proposal, following the instructions for requesting a supplement to an existing NSF award. Your Supplementary Document can be up to three pages (as in the request for a supplement to an existing NSF award) and should address the components described above (a description of the REU activities, etc.). The REU Solicitation also provides guidance on how to incorporate the REU budget into the budget of your full proposal.

2.10 RESEARCH IN UNDERGRADUATE INSTITUTIONS

NSF provides a special opportunity for faculty at predominantly undergraduate institutions to engage in research through the Research in Undergraduate Institutions (RUI) program. (NSF's definition of predominantly undergraduate institution is found in the RUI Solicitation; see also Chapter 6, Section 6.6, for the Carnegie Classification.) The RUI program offers three avenues for funding: (1) funding for independent research by a faculty member or collaborative research by groups of faculty members, (2) funding for the purchase of shared-use instrumentation, and (3) Research Opportunity Awards to enable faculty members at predominantly undergraduate institutions to engage in research as visiting scientists with NSF-supported PIs at other institutions.

Proposals submitted to the RUI program are prepared as you would for regular NSF programs except that you must include an RUI Impact Statement describing the anticipated effects of the research on the institution and a Certification of RUI Eligibility. RUI proposals will be marked as such (with "RUI:" before the title of the proposal) but will be reviewed along with other proposals (RUI and non-RUI) in the same area of research (and the funds come from the same pot of money). However, the reviewers are given special instructions on how to review the Impact Statements and the special circumstances for the proposed RUI projects. Thus, when you prepare an RUI proposal, you should take advantage of the RUI nature and capitalize on the resources, student training opportunities, and curriculum innovations at your institution, so that you can highlight the strengths of your proposal as an RUI proposal distinct from proposals from other research institutions.

2.11 OTHER SPECIAL TYPES OF PROPOSALS

As described in Chapter 1, Section 1.3, NSF provides special opportunities for projects that have extraordinary urgency with regard to time (Grants for Rapid Response Research, RAPID) or to support exploratory but potentially transformative work (Early-Concept Grants for Exploratory Research, or EAGER). A note of caution: although we discuss both RAPID and EAGER grants together in this section, they are used for very different purposes. Generally, a proposal will fit into either the RAPID or the EAGER category; these two funding mechanisms are not necessarily interchangeable.

Before you submit a RAPID or EAGER proposal, it is imperative that you talk to an NSF Program Officer about your idea to make sure that your idea is viable for one of these programs. NSF Program Officers can provide helpful guidance so that you do not waste time preparing a proposal that might not be a good fit for either the RAPID or the EAGER program. Also, once a Program Officer gives you the go-ahead, that Program Officer will know to look out for your proposal once it is submitted through FastLane. For both EAGER and RAPID proposals, you will prepare a proposal with the following guidelines (consult also the GPG, Chapter II, Section D):

- The Project Description should be brief (two to five pages for RAPID and five to eight pages for EAGER). For RAPID proposals, you must explain why the proposed research is urgent and why a RAPID award is the most appropriate funding mechanism for the work. For EAGER proposals, you must justify why the research is a good fit for EAGER.
- The box for either RAPID or EAGER must be checked on the Cover Sheet when you submit your proposal (see Chapter 3 for a discussion about submitting proposals).
- You can request up to $200,000 and a duration of one year for RAPID grants; you can request up to $300,000 and a duration of up to two years for EAGER grants.

Typically, both RAPID and EAGER grants are reviewed internally. As such, the processing time is much shorter than with proposals to regular programs.

2.12 HUMAN AND VERTEBRATE ANIMAL SUBJECTS

If your research proposal involves human or animal subjects, you will need to check the appropriate box on the Proposal Cover Page. Research involving human subjects must ensure that participants are protected from research risks in compliance with federal policy. This means that your project must be reviewed by your institution's Institutional Review Board (IRB) and be granted IRB approval to conduct the research with human subjects or be declared exempt from IRB review. You may choose to submit an application for IRB review around the time that you submit your proposal to NSF, or you may choose to wait until you hear from NSF. However, know that NSF will not be able to make an award until you have IRB approval or a declaration of exemption (see also the discussion of IRB in Chapter 6, Section 6.1, when it comes to the funding stage).

Research involving vertebrate animal subjects must comply with the Animal Welfare Act and federal regulations pertaining to the care, handling, and treatment of vertebrate animals for research or educational purposes. Similar to research with human subjects, if you propose to conduct research with vertebrate animal subjects, you must receive approval from your institution's Institutional Animal Care and Use Committee (IACUC) before NSF can make an award.

Chapter 3

Submitting Your Proposal

3.1 GETTING READY TO SUBMIT

Your proposal is written, your budget is finalized, you've collected all the information you need, and now it is time to upload your documents for review. Each institution may have a different set of rules governing the actual submission process, so you need to check with your Sponsored Research Office (SRO) to make sure that everything is in order. For example, there may be several certification pages that you need to sign, some of which are internal forms and others mandated by federal agencies.

If the program to which you are submitting has a strict deadline, then you have until 5 p.m. (your local time) on the day of submission to submit your proposal (after which time the NSF system will simply not accept submissions; see also Chapter 2). If you are submitting to a regular program (see Chapter 1, Sections 1.2.3 and 1.3.2), you may not have a hard deadline but rather a so-called target date. If you can adhere to the target date that would be great, but if you need a few more hours past the target date, that is not a big deal. Sometimes the Program Officer may give PIs a few more days past the target date; all you need to do is ask, and you may get permission to submit late (it's nice to have some NSF flexibility!).

3.2 SUBMITTING PROPOSAL DOCUMENTS

You can submit proposals to NSF in two ways (both electronically): either through FastLane (http://fastlane.nsf.gov/), NSF's online proposal submission and review system, or through Grants.gov, the government-wide portal for finding and applying for federal grants online. These two electronic systems are completely separate. That means, for example, if you regularly use FastLane but have never used Grants.gov, your FastLane login information will not work on Grants.gov. We talk about FastLane and Grants.gov in turn.

Most proposals to NSF are submitted through the FastLane system. Because FastLane is internal to NSF, the process of getting your proposal to the program to

Having Success with NSF: A Practical Guide, First Edition. Ping Li and Karen Marrongelle.
© 2013 Wiley-Blackwell. Published 2013 by John Wiley & Sons, Inc.

which you have submitted is more streamlined than with Grants.gov, and the Fast-Lane system is specific to NSF requirements and proposal language. For instance, FastLane technical support persons are housed in the same building as the Program Officers. Should your grant be funded, you will need to manage your grant through the FastLane system, so you will get to know FastLane at some point along the way!

No matter which system you use, you should spend some time getting familiar with either FastLane or Grants.gov well before the planned date of your submission. Prior to the first time logging into FastLane or Grants.gov, you will need to obtain an NSF ID from your SRO (you can also email fastlane@nsf.gov for this; see Section 3.2.2 below). We recommend, prior to your first grant proposal submission, that you take an hour or so to navigate through the FastLane or Grants.gov submission system. When you are familiar with how to navigate in one of the online submission systems, uploading the components of your proposal will be all the smoother.

3.2.1 Submission through Grants.gov

Grants.gov is a clearinghouse and resource for finding and applying for funding across federal agencies in the United States. One feature of Grants.gov is that you can search for and apply for federal grants in one place. Grants.gov was established in 2002, under President Bush's Fiscal Year Management Agenda, to improve government services to the public. All NSF Program Solicitations are listed on Grants.gov. You can search Grants.gov by agency, funding activity category, eligibility type, and dates, among other criteria. Grants.gov is an excellent resource to find funding opportunities in your area beyond NSF. NSF publishes a Guide for Preparation and Submission of NSF Applications via Grants.gov, which can be found at: http://www.nsf.gov/pubs/policydocs/grantsgovguide0113.pdf?WT.mc_id=USNSF_109. Because most grant proposals are submitted to NSF through FastLane, we will focus the remainder of this chapter on FastLane-related submission issues.

3.2.2 Submission through FastLane

NSF's FastLane electronic system can be accessed at http://fastlane.nsf.gov. It is a comprehensive submission, review, and reporting system used at NSF. The FastLane Helpdesk is a very useful resource if you are having difficulty navigating in FastLane or uploading documents, or if you experience any other FastLane-related problems. You can also send an email to fastlane@nsf.gov during off-business hours, and they are usually very responsive. In the email, you specify your problem clearly, with a relevant subject line. If you can provide screenshot of your problem, that would be even more helpful to the FastLane technical support personnel.

The FastLane homepage contains Advisory Warnings, Quick Links, User Support Contact, and Navigation tools for the Fastlane system. Figure 3.1 provides a screen shot of the homepage.

To upload a proposal to FastLane, click on the "Proposals, Awards, and Status" link on the menu bar toward the top of the page. This is the link you will use to

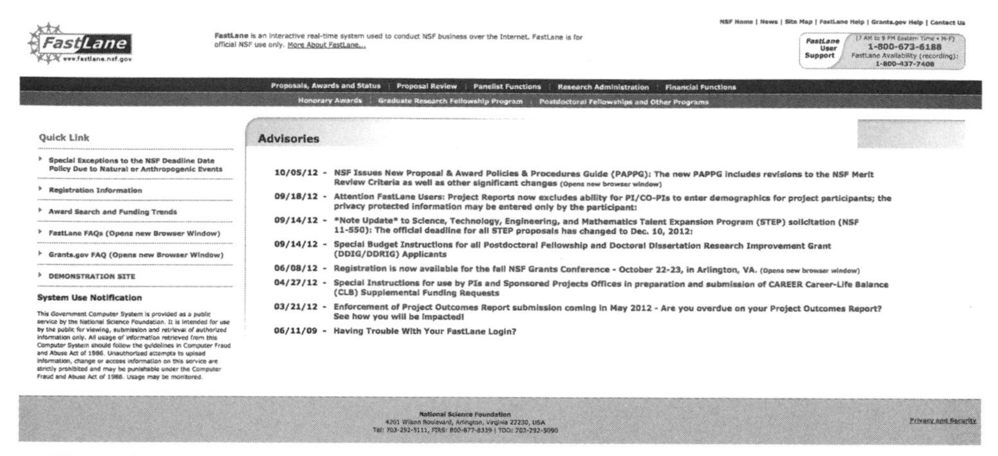

Figure 3.1. FastLane Homepage screen shot (courtesy of the National Science Foundation).

upload your various proposal documents and to check on the status of your proposal while it is under review. The other tabs are used for proposal review and research administration, accessible only to reviewers, panelists, and Program Officers. There are also two tabs reserved for the Graduate Research Fellowship Program and the Postdoctoral Fellowships under the main FastLane functions.

You will log onto FastLane as a PI/co-PI by entering your last name, NSF ID, and a password into the login box. If you have not used FastLane for a while and forgot about your NSF ID, you can send a request to fastlane@nsf.gov, specifying your name, institution, and email address (see also above). FastLane will respond to your request and send your NSF ID. Once you have your NSF ID, you can also reset your password online by following the link under "PI/Co-PI Log In" at https://www.fastlane.nsf.gov/jsp/homepage/proposals.jsp.

To begin uploading your documents into Fastlane, you will choose the "Proposal Functions" link from the menu under the heading "What Do You Want to Work On?" You will then be able to choose from a number of submission options, including Uploading Letters of Intent, Proposal Preparation, and Proposal File Update. Figure 3.2 shows a screen shot of the PI management page once you are logged in. Figure 3.3 shows a screen shot of the page you get when you click on Proposal Functions.

3.2.2.1 Submitting Letters of Intent

When you click on the Letters of Intent link, you will be directed to a page listing available Program Solicitations for which Letters of Intent (LOIs) can be submitted. If you are trying to submit a LOI for a Program Solicitation that is not listed, contact the FastLane Helpdesk at 1-800-673-6188 (this number is also posted on the FastLane homepage; see FastLane User Support numbers at the right-hand

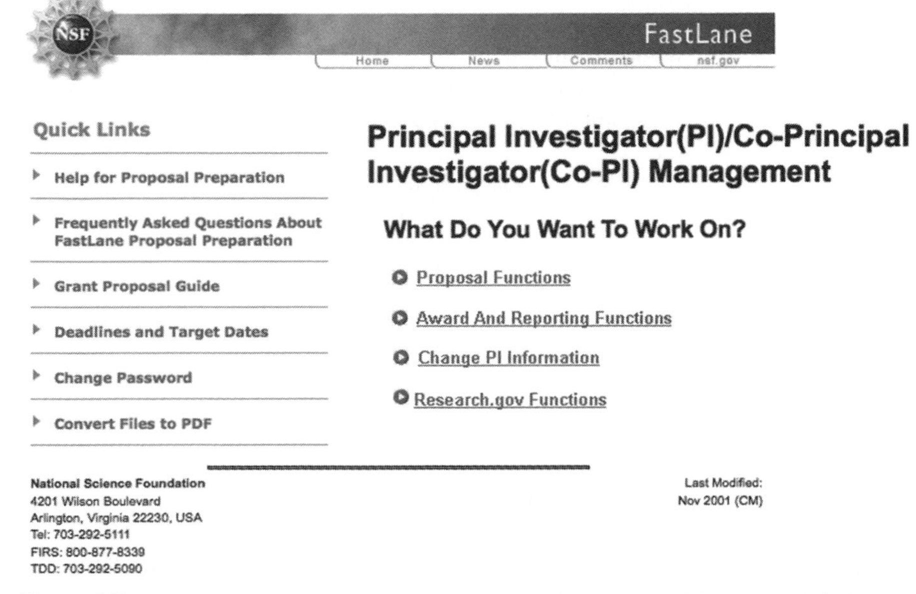

Figure 3.2. PI Management Menu screen shot (courtesy of the National Science Foundation).

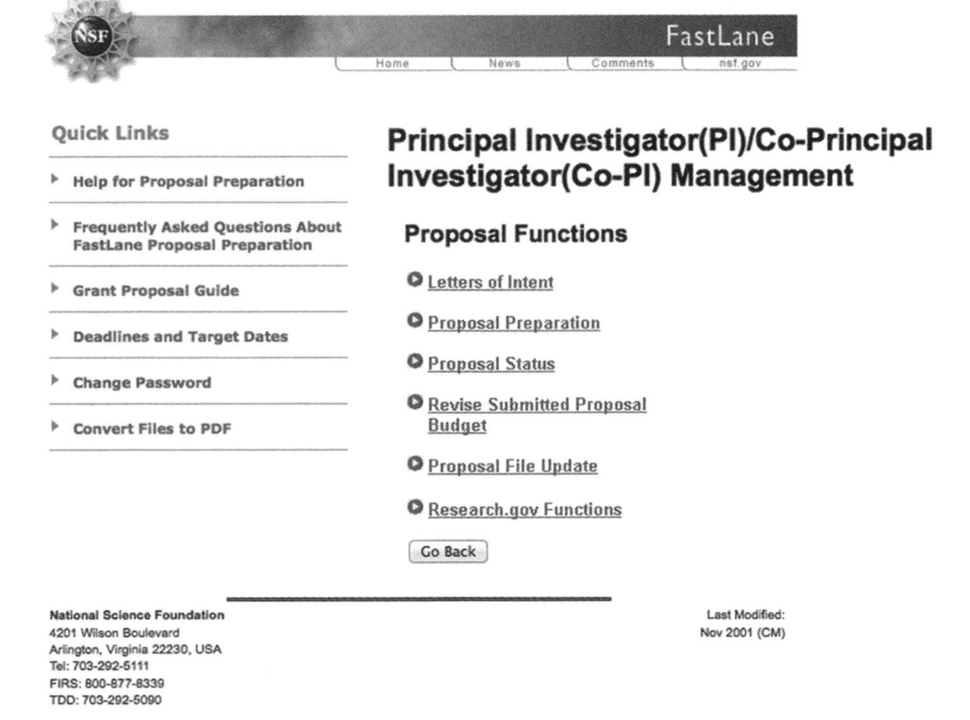

Figure 3.3. Proposal Functions menu screen shot (courtesy of the National Science Foundation).

corner) or email fastlane@nsf.gov. Remember that not all Program Solicitations request a Letter of Intent, and not all Letters of Intent are required. NSF Program Officers use the information from Letters of Intent to begin to form review panels and line up ad hoc reviewers. The information you provide in an LOI helps ensure that the most appropriate and expert reviewers evaluate your proposal (see also Chapter 2, Section 2.3.1).

When you locate the Program Solicitation for which you want to upload your Letter of Intent, click on the "Create" button under the LOI Action column. You then will be directed to a screen where you will input:

- Project Title
- A synopsis of the project in 2,500 (maximum) characters
- Point of Contact for NSF Inquiries
- Project PI

Depending on the Program Solicitation to which you are responding, you might be asked to include additional information such as partner institutions, your institution type, the type of project you are submitting (if there are strands or tracks in your Program Solicitation), or other senior project personnel (see an example in Chapter 2, Figure 2.1).

3.2.2.2 Submitting Full Proposals

To submit the documents that comprise your full proposal, click on the Proposal Preparation link under the Proposal Functions menu (see Figure 3.3).

When you initially enter the Proposal Preparation area, you will be taken to a screen with your PI information. If you need to update or edit any of your information, you can do that by clicking on the "Edit PI Information" button. It is very important that you do so if you have changed institutions since you last used FastLane. Some of the PI information will appear on the cover page automatically, and some other information is used internally only by NSF (e.g., for statistics such as PI's demographic and geographic information). To begin uploading your proposal documents, click on the "Prepare Proposal" button. You will be directed to a screen where you can create a new proposal.

Initially, you will click on the "Create Blank Proposal" button. Once you have uploaded a proposal, you may want to create a template that saves information from an already uploaded proposal to use in subsequent proposal submissions. Small Business Innovation Research (SBIR) and Small Business Technology Transfer Program (STTR) each have their own proposal forms.

When you click on the "Create Blank Proposal" button, two important things happen: you will (1) be automatically assigned a Temporary Proposal number and (2) be taken to the Form Preparation page, where you can upload your proposal documents. If someone else is helping you submit your proposal, such as your SRO or a Research Assistant, you will need to provide them with the Temporary Proposal number (and the title) so that they know which proposal to access. This is important especially if you are working on multiple proposals at the same time. At this point, you can upload the documents that you've prepared (See Chapter 2, Section 2.4). FastLane will automatically create a Table of Contents for your proposal, so this is one piece that you don't have to worry about! Figure 3.4 presents a screen shot of the proposal preparation form. Simply click on the GO button each time and follow the instructions to upload documents.

You should take care when completing the Cover Sheet. There are four sections to the Cover Sheet: (1) Awardee Organization and Primary Place of Performance, (2) Program Announcement/Solicitation Number, (3) NSF Unit Consideration, and (4) Remainder of the Cover Sheet.

The Awardee Organization information should be automatically populated for you. If you do not see this information correctly, check with your SRO. You will need to enter the information about the Primary Place of Performance, whether or not it is the same information as the Awardee Organization. For instance, you may work at a research center that has a different location than the Awardee Organization (off campus, for instance), where the work will be conducted. Alternatively, you may be proposing to conduct research internationally, at a site different from your home institution. You will want to update the Primary Place of Performance information in these types of cases to list the address of the research center where the work will be conducted or the location of the international work.

You must choose one (and only one) Program Announcement/Solicitation number or you will choose to submit your proposal to the Grant Proposal Guide (GPG). Typically, you will select to submit your proposal to the GPG when you are submitting a proposal to the RAPID or EAGER program (see Chapter 2, Section 2.11) or if your Program Officer directs you to do so. The Program Announcement/ Solicitation numbers are listed in reverse order, starting with the most recently released Program Announcement/Solicitation first. The Program Announcement/ Solicitation numbers begin with the fiscal year in which the Program Announcement/ Solicitation is released. For instance, Program Announcement 12-566, Focused Research Groups in the Mathematical Sciences Program, was released in the fiscal year 2012 and given unique Program Announcement number 566.

Once you've chosen a Program Announcement/Solicitation number, you will need to select an NSF unit under which your proposal will be reviewed. In most cases, the NSF unit selection will be straightforward because for many Program Announcements/Solicitations, one division administers the program (see Chapter 1, Section 1.2.3, for determining which program to select). In cases where more than one division or directorate administer a program, you will need to select the division or directorate that best matches your area of study. For instance, the NSF Early CAREER Program is administered foundation-wide. Once you select the CAREER Program Announcement number, you will be prompted to select the division and/or program that most closely relates to your proposed project. You should contact a Program Officer if you have questions about which unit to associate with your proposal, particularly if you are submitting a CAREER, EAGER, or RAPID proposal.

If you feel that your proposal fits more than one program at NSF, you can list multiple programs, but one program must be identified as the primary program and

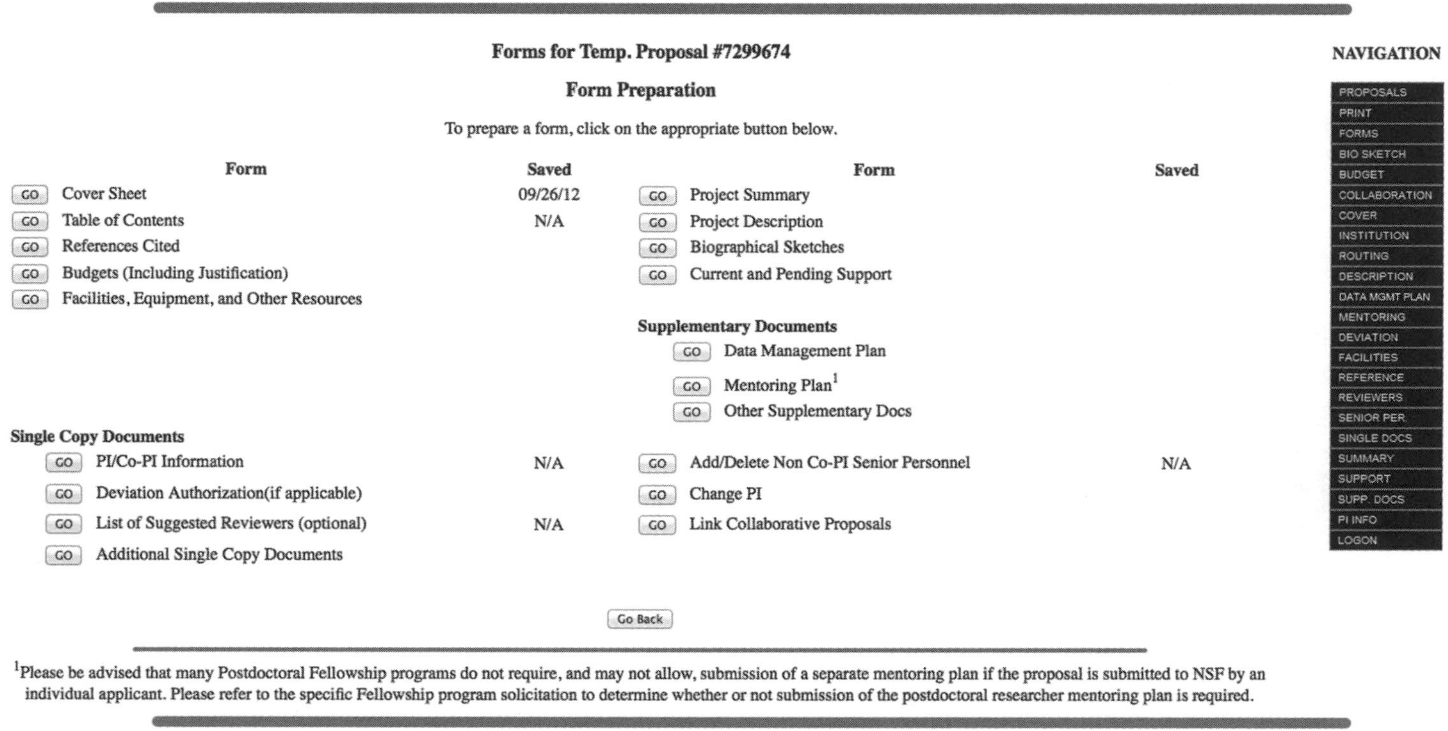

Forms for Temp. Proposal #7299674

Form Preparation

To prepare a form, click on the appropriate button below.

Form	Saved	Form	Saved
GO Cover Sheet	09/26/12	GO Project Summary	
GO Table of Contents	N/A	GO Project Description	
GO References Cited		GO Biographical Sketches	
GO Budgets (Including Justification)		GO Current and Pending Support	
GO Facilities, Equipment, and Other Resources			
		Supplementary Documents	
		GO Data Management Plan	
		GO Mentoring Plan[1]	
		GO Other Supplementary Docs	
Single Copy Documents			
GO PI/Co-PI Information	N/A	GO Add/Delete Non Co-PI Senior Personnel	N/A
GO Deviation Authorization(if applicable)		GO Change PI	
GO List of Suggested Reviewers (optional)	N/A	GO Link Collaborative Proposals	
GO Additional Single Copy Documents			

Go Back

[1]Please be advised that many Postdoctoral Fellowship programs do not require, and may not allow, submission of a separate mentoring plan if the proposal is submitted to NSF by an individual applicant. Please refer to the specific Fellowship program solicitation to determine whether or not submission of the postdoctoral researcher mentoring plan is required.

NAVIGATION

PROPOSALS
PRINT
FORMS
BIO SKETCH
BUDGET
COLLABORATION
COVER
INSTITUTION
ROUTING
DESCRIPTION
DATA MGMT PLAN
MENTORING
DEVIATION
FACILITIES
REFERENCE
REVIEWERS
SENIOR PER.
SINGLE DOCS
SUMMARY
SUPPORT
SUPP. DOCS
PI INFO
LOGON

Figure 3.4. Proposal Preparation Form screen shot (courtesy of the National Science Foundation).

others as secondary or tertiary. In Chapter 4 (see Section 4.3.3), we discuss the co-review process, and the advantages and disadvantages of having your proposal reviewed by more than one program.

The remainder of the Cover Sheet asks for the following:

- The title of your proposed project
- The duration (in months) of your project and expected start date (note: FastLane will automatically populate your total budget request on the Cover Page for you, once you have completed the budget pages)
- Information about co-PIs
- Information about previous awards for which the proposal is a renewal or has an associated preliminary proposal (this is uncommon)
- Whether you are submitting your proposal simultaneously to another federal agency (see Chapter 1, Section 1.2.1, for an example)
- Whether you or your proposal falls into any special categories (e.g. EAGER or RAPID grant; your research involves human subjects; or you previously have not been awarded an NSF grant).

It is a good idea to upload your documents in stages and budget time for uploading problems. During peak submission windows for a program that receives thousands of proposals, for instance, the system can be slow, so it is best not to wait until 4:30 p.m. local time of the deadline to begin uploading your documents. Your institution may have deadlines weeks earlier than the date that the proposal is due into NSF in order to give your SRO time to review the proposal—and give you time to respond to any issues—before it is submitted. In Chapter 2 (see Section 2.4), we discussed the various components/documents required of a full proposal by NSF, including biographic sketches, budget and budget justification, current and pending support forms, and other supplementary documents in addition to the main contents of the proposal. These should all be carefully prepared for each PI and co-PI (and in some cases consultants) ahead of the actual submission process.

At some point during your proposal uploading process, you will need to give your SRO access to your proposal. Just as you have menus and options as a PI in the NSF FastLane system, your SRO also has a variety of different menus and options. When you click on the "Allow SRO Access" button under Proposal Actions, you will have the option of allowing your SRO to (1) only view the proposal, (2) view and edit the proposal, or (3) view, edit, and submit the proposal. You should work with your SRO to determine when you should allow which type of access. For instance, if your SRO will upload your budget, then you will want to give access to at least view and edit your proposal early on. Some SRO offices may want to ask for option 3 from the beginning so that they can check and correct any issues as they arise before submitting. It is your SRO who actually submits your proposal to NSF—your SRO pushes the proverbial "Submit" button. This is because it is your institution, not you as an individual, who submits proposals to NSF (see Chapter 1, Section 1.2.2, for discussion of the NSF–awardee relationship). It is a good idea to communicate regularly with your SRO before and as you are uploading documents to submit to NSF.

Before you submit your proposal, you or your SRO should use the FastLane "Check" button feature to verify that your proposal is complete. By clicking the "Check" button under the Proposal Actions menu and selecting the appropriate Temporary Proposal number, FastLane will scan your documents and tell you which ones are missing and of those, which are required for review. If you submit a proposal without some required documentation, NSF can return your proposal without review. In this regard, it is also important to follow the NSF guidelines as specified in the Grant Proposal Guide (GPG, http://www.nsf.gov/publications/pub_summ.jsp?ods_key=gpg) with regard to the specific requirements on not only the contents of the proposal but also the style including font size, margin, and spacing of letters. Your SRO staff can give you valuable information on these details.

3.2.2.3 Proposal PIN

If you need to allow individuals other than your SRO access to your proposal, for instance if you are submitting a Collaborative Research proposal or if you want a Research Assistant to help upload documents, you will need to create a proposal pin for your proposal. By clicking on the "Proposal Pin" button under the Proposal Actions menu, you will be prompted to enter a four-digit code. Remember this number! You can then give this number to others who will need it to access your proposal.

3.2.2.4 Correcting Errors

Occasionally you or your Program Office will notice something in error about your proposal after it has been submitted. NSF may allow you to fix the problem before the proposal goes out to reviewers. If you need to fix part of an already submitted proposal, you will need to create a Proposal File Update and provide a justification that addresses why the changes are being requested and outlining the differences between the original and replacement files. Unless your Proposal File Update is submitted before the deadline in a Program Solicitation or before external review (in cases with target dates), the Proposal File Update will need to be approved by your Program Officer, so you will want to talk with your Program Officer before submitting a Proposal File Update. The Proposal File Update can be found under the Proposal Functions page (see Figure 3.3). Of course, it is best to check for all possible errors before submission, and the following Checklist will help you in this regard.

CHECKLIST FOR NSF PROPOSAL SUBMISSION

The Grant Proposal Guide (GPG) has a more detailed Proposal Preparation Checklist (Exhibit II-1) that you should consult (see http://www.nsf.gov/pubs/policydocs/pappguide/nsf13001/gpg_2.jsp#IIex1). The following is an abridged checklist for the major components of your proposal before you submit.

☐ **Formatting Requirements**

Use correct fonts and margins in the text: Arial, Courier New, or Palatino Linotype at 10 points or larger; Times New Roman at 11 points or larger; one inch margins, in all directions.

See *Grant Proposal Guide*, Chapter II, Section B: Format of the Proposal, for details.

☐ **Project Summary** (1 page)

Include Overview, Intellectual Merit, and Broader Impacts paragraphs.

☐ **Project Description** (15 pages)

Include discussion of Intellectual Merit and Broader Impacts. Include a separate section with discussion of the Broader Impacts. See this book, Chapters 2, 4, 5.

☐ **References Cited** (no page limits)

Follow accepted scholarly practices in providing citations.

☐ **Budget and Budget Justification** (Budget Justification includes no more than 3 pages for all years)

☐ **Facilities** (no page limits)

☐ **Biographical Sketches** (for all senior personnel; 2 pages per person)

☐ **Current and Pending Support** (for all funded projects you are working on, including the proposal you are submitting)

☐ **Supplementary Documents**

Postdoctoral Researcher Mentoring Plan (1 page)

Data Management Plan (2 pages)

Letters of Collaboration(if applicable, no page limits)

Letters of Support(if applicable; no page limits)

For all the above except Formatting, see *Grant Proposal Guide*, Chapter II, Section C: Proposal Contents, and this book, Chapter 2, Section 2.4, for details.

☐ **Provide Other Institution Contact to Sponsored Research Office** (for Collaborative Proposals only)

See *Grant Proposal Guide*, Chapter II, Section D: Special Guidelines, for details.

☐ **Allow Sponsored Research Office full access to proposal inside FastLane when ready to submit**

Chapter 4

Reviewing of Your Proposal

4.1 INTRODUCTION

It usually takes several months of hard work to get a regular research proposal written, so the PI feels very relieved when the submission button is hit (or more precisely, when told by the SRO that the final submission button has been hit). Most people want to take a break and simply not worry about their proposal until a few months later when they start to get anxious about the outcome of the review and the decision from NSF. There is also not much they can do during this period of waiting (see Fig. 1.2 in Chapter 1 for the length of this period and the relevant process). Nonetheless, it might be helpful for researchers to have some knowledge about what exactly goes on at NSF during this period, so that things can be put into perspective when the review results do come. This knowledge might also be very important for later stages involving revision and resubmission in case the review results are less than completely positive (see Chapters 5 through 7). Providing this knowledge is the purpose of this chapter.

Perhaps all PIs have had the experience of submitting a manuscript to a journal before they have had the experience of submitting a grant proposal to NSF or other funding agencies. It might be useful to draw some quick comparisons here between the review of a journal article and the review of a grant proposal. For journal articles, the process is usually quite simple, in that most journals have the so-called peer review process, which works something like the following: the editor or editors quickly look at your submission to determine its appropriateness for their journal and assign your submission to an action editor who oversees the review process, and then the action editor sends the manuscript out to experts in the field for comments (or rejects your manuscript right away if the editor feels that your manuscript does not meet the journal's standard). After a few months (depending on the journal), the editor receives comments from the expert reviewers, makes a judgment on the merit of your article on the basis of his or her own reading and the reviewers' comments, and reaches a positive or negative decision on your submission.

For grant proposals, the process is similar, but there are also other aspects involved. Some of these other things may be of no interest to the researcher at first glance, but they may turn out to be crucial (or fatal!) in some cases. For example,

Having Success with NSF: A Practical Guide, First Edition. Ping Li and Karen Marrongelle.
© 2013 Wiley-Blackwell. Published 2013 by John Wiley & Sons, Inc.

at NSF, before your proposal even reaches the Program Officer, the program assistant checks it for compliance, with regard to whether your proposal is prepared according to the guidelines specified in the Grant Proposal Guide (see Chapter 2 for some discussion of this guide). The compliance check can be quite specific, for example, on the font size and spacing used for the text, and on whether your proposal includes discussion of both intellectual merit and broader impacts (two NSF review criteria) in the text and in the Project Summary (there are new NSF guidelines regarding how Project Summary and Project Description should be structured; see Chapter 2, Sections 2.4.2 and 2.4.3). If your proposal is found to be noncompliant, it may be returned to you without review. The Program Officer can also return the proposal without review if your proposal is found to be irrelevant to the mission and the focus of the program; for example, if your project is primarily interested in clinical diagnosis or treatment of behavior, it would be more suitable for NIH than for NSF (see discussion in Chapter 1, Section 1.2.1, regarding NSF vs. NIH). This is why it is important to check with the relevant Program Officers before you submit if you have any doubt about the relevance and appropriateness of your proposal to the program (see Chapter 1, Section 1.3.1, for discussion of how to contact the Program Officer). Only after your proposal is deemed relevant and compliant does the review process begin.

Increasingly, some academic journals are also doing much of what is common at NSF, given the large number of submissions they receive these days: articles are returned to authors without review because they do not conform to the style requirement, do not meet the quality standards, or do not match with the aims and scopes of the journal. What really differentiates journal reviews and NSF reviews lies in the specific criteria with which the review is conducted.

For NSF, the two review criteria used to evaluate proposals are Intellectual Merit and Broader Impacts (see NSF's updated Merit Review website, along with Merit Review Facts at http://www.nsf.gov/bfa/dias/policy/merit_review/; see also Chapter 2, Section 2.5). First, for Intellectual Merit, reviewers are asked to comment on the potential of the proposed activities to advance knowledge, and for Broader Impacts, reviewers are asked to comment on the potential of the project with regard to its benefit to society and to achieving specific, desired societal outcomes. NSF has renamed its review criteria as the "Merit Review Principles and Criteria," and reviewers are asked to use three guiding principles, two review criteria, and five review elements (see Chapter 2, Section 2.5). To reiterate, the three guiding review principles are: (1) All NSF projects should be of the highest quality and have the potential to advance, if not transform, the frontiers of knowledge; (2) NSF projects, in the aggregate, should contribute more broadly to achieving societal goals; and (3) assessment of projects should be based on appropriate metrics, considering the resources available to the investigators and the corresponding effects of Broader Impacts. Application of the last principle will depend on the size of the project, which is done often at an aggregated or large-scale level. Second, the two review criteria remain Intellectual Merit and Broader Impacts, but reviewers now need to ask the what, why, and how questions: what the proposers want to do, why and how they plan to do it, how they know if it will succeed, and what benefits will accrue if the project is successful. These questions will be applied to both review criteria. With regard to Broader Impacts, NSF has previously provided a list of specific set of activities to researchers (e.g., outreach to the public or K-12 students), but it now leaves the details of such activities to the PI to decide (in light of program-specific priorities and NSF core missions), except that it continues to consider that broadening the participation of underrepresented groups is a critical component of Broader Impacts as well as an NSF priority. Finally, the five review elements provide further contexts for applying the two review criteria to proposals: (1) What is the potential for the proposed activity to advance knowledge and understanding within and across research fields, and to benefit society or advance desired societal outcomes? (2) To what extent do the proposed activities suggest and explore creative, original, or potentially transformative concepts? (3) Is the plan for carrying out the proposed activities well reasoned, well organized, and based on sound rationale? Does the plan incorporate a mechanism to assess success? (4) How well qualified is the individual, team, or institution to conduct the proposed activities? and (5) Are there adequate resources available to the PI (either at the home institution or through collaborations) to carry out the proposed activities?

4.2 WHO'S WHO IN THE REVIEW PROCESS

The core value of the peer review process is the same for grant proposals as for journal articles (and indeed, for most academic output), in that expert opinions are sought in a fair and competitive fashion to allow for an informed decision on the support or publication of a product under consideration in light of its intellectual merit or quality. NSF regards expert comments highly, and has developed a rigorous review system to include multiple perspectives from different sources. Figure 4.1 shows a diagrammatic sketch of who's who in the review process.

The Program Officer initiates the process by looking at the pool of NSF reviewers, inviting ad hoc reviewers (experts who have more specific knowledge in your domain of research) and forming panels (consisting of 6–15 members who have more general knowledge of the field and are able to comment on proposals not directly in their own areas). The process ends with the Program Officer who analyzes the comments of the reviewers and panelists and makes a final recommendation on a proposal.

4.2.1 Panelists

Shortly after your proposal arrives at the desk of the Program Officer, a panel (formally the "Advisory Committee") is convened. This panel consists of scholars who are experts in a domain of research. The word "domain" here should be interpreted broadly to refer to a large research area or several areas covered by the NSF program, such as digital resources, cognitive neuroscience, or experimental physics. When inviting the panel members (or "panelists"), the Program Officer bears in mind the relevance of a potential panelist's research, but more importantly, tends to invite those who possess a broad knowledge base of the science and engineering subfields

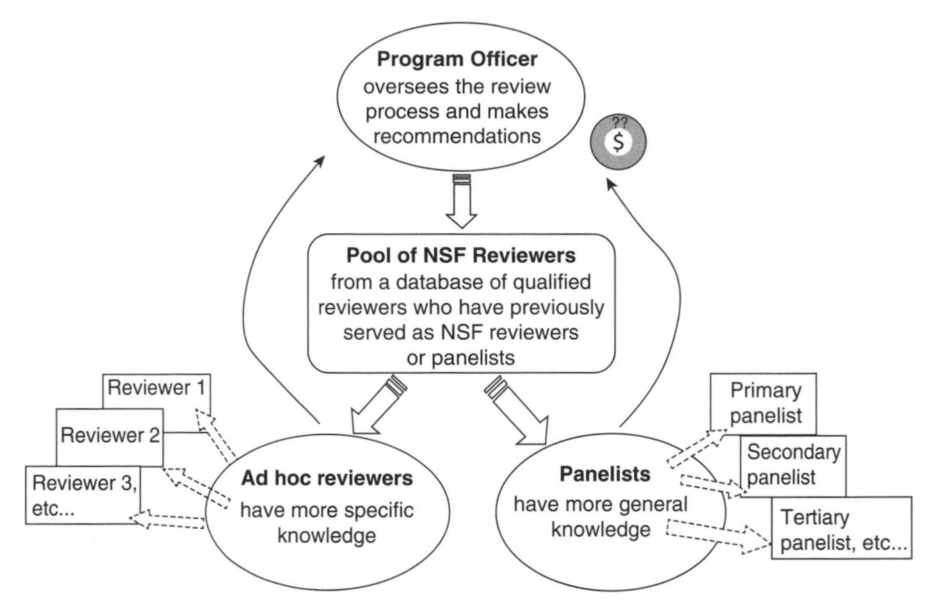

Figure 4.1. Who's who in the review process: The Program Officer initiates the process by looking at the pool of NSF reviewers, inviting ad hoc reviewers (experts who have more specific knowledge in your domain of research) and forming panels (consisting of people who have more general knowledge of the field and are able to comment on proposals not directly in their own areas). The process also ends with the Program Officer, who analyzes the comments of the reviewers and panelists and makes a final recommendation on a proposal.

represented by the research community of the program. The implication of this is that the panelist may not necessarily be an expert in the specific domain of the proposed research, but is knowledgeable enough to judge the scientific merit and methodology of the proposed project. Because the scope of an NSF program is usually quite broad, it would be unimaginable to convene a panel that has experts in every small domain as demanded by the number of proposals to be reviewed. What all this suggests is that you as an investigator should write a proposal not just for someone exactly in your field, as you would in writing an empirical article, but a proposal clear enough for a "generalist" expert on the panel (see Chapters 2 and 5 for more on this). This is also important as NSF puts more emphasis on "transformative research," and panelists tend to look beyond the specific research paradigm for broader impacts and significance.

In addition to considering the representation of research domains against the proposals to be reviewed, the Program Officer also considers the issue of panel diversity when inviting and selecting panelists. Diversity of a panel is reflected on, other things being equal, the type of institutions the panelist represents (e.g., research university vs. smaller colleges), geographic location of the panelist's institution (e.g., East Coast, West Coast, Midwest), career stage of the panelist (senior or mid-career researchers vs. early-career researchers), and ethnicity and minority status of the panel members. NSF encourages programs to build a balanced portfolio, including the use of reviewers and panelists, so that funding decisions will not be biased, for example, toward only big research schools, specific geographic regions, or seasoned scholars. Other federal funding agencies (e.g., NIH) also have similar criteria or considerations when they invite panelists or reviewers.

Typically, a panelist reviews about 10–20 proposals for a given panel meeting, although this varies from program to program, and from time to time (e.g., the fall panel may have more proposals to review than the spring panel, probably because people have more time to prepare proposals in the summer). To take a simplified situation, let's imagine a typical panel that consists 14 panelists who need to review 100 proposals. If each proposal needs to receive two reviews, the Program Officer needs to assign each panelist 14 proposals (but four panelists will need to review 15 proposals). Before the Program Officer assigns the proposals to the panelists, he or she might ask the panelists to first indicate, for each proposal, the panelist's preference of review (e.g., on a 1–5 scale). In the end, the Program Officer will have to match proposals with panelists, and naturally, some proposals will be closely matched to the panelist's research expertise, while others may only be loosely matched (in some cases the Program Officer may do the match without first consulting the panelists). Panelists are always asked to take a broader perspective than just their own research in evaluating the scientific merit and broader impacts of proposals.

Unlike the NIH, which publishes a roster of the members of a panel (Study Section), NSF keeps the identity of the panel members confidential, and thus you cannot ask the Program Officer who are on a panel. Indeed, NSF requests that panelists share no information about the panel or the review process with anyone outside the panel.

4.2.2 Ad Hoc Reviewers

Most, but not all, regular NSF programs use ad hoc reviewers in addition to panelists to help with the review process, which is again different from the NIH review process (no ad hoc reviewers are used for NIH proposals). Some NSF programs, especially one-time, time-sensitive, or large-scale programs, may involve only panelists and no ad hoc reviewers. In such cases, there may be more panelists involved in each panel, so that a minimum number of reviews can be obtained for each proposal—a number that could range from 3 to 5, depending on the panel and program (you may get more than five reviews, if your proposal is co-reviewed by two or more programs; see Section 4.3.3).

Unlike the panelists who will be asked to review many proposals for a given panel, ad hoc reviewers are sought by NSF Program Officers to comment on one or a few specific proposals that fall squarely within the reviewer's own research area. Ad hoc reviewers may not play a decisive role (but see below) in recommending a proposal for funding because they only provide written comments and are not present at the panel to voice their opinions. However, the addition of the ad hoc reviews to the system is important, particularly in light of the fact that some of the panelists may not be directly working in the same area as the proposal's investigators, or may not be familiar with the specific research paradigms or methods. Ad hoc reviewers often provide very detailed, organized, and insightful comments that not only help the panel and the Program Officer to make recommendations, but also help the investigators to improve the quality of the project and quality of the proposal. The panel and the Program Officer take ad hoc reviews very seriously, especially when the ad hoc reviewer is a known expert in the field and provides pertinent comments. It is important to note, however, that the goal of NSF panelists and reviewers is not to help the investigator to improve the research, but rather to provide expert comments on the merits of the proposed project so that NSF can make an informed decision. This goal is sometimes different from that in the journal peer reviewing process, although in practice many NSF reviewers do provide constructive and helpful comments that eventually lead to better quality of the proposed research. In any case, both the panelists and ad hoc reviewers are asked to follow the same set of three guiding principles, two review criteria, and five review elements to evaluate research proposals (see Chapter 2, Section 2.5; and this chapter, Section 4.1).

Given the importance of ad hoc reviews for NSF, it is your job to suggest as many good ad hoc reviewers as possible under the List of Suggested Reviewers, an option provided to investigators by NSF. It would be a mistake not to use this option, but in suggesting reviewers you should exclude people with conflicts of interests, such as your collaborators, co-authors, or former advisors or mentors. The Program Officer does not know every qualified reviewer for every field out there, and therefore he or she can use some of your suggestions. Sometimes the reviewer you suggest may even become a panelist down the road if the reviewer provides great comments and the Program Officer gets to know the person well. Just as with a journal article review, the final selection of reviewers could be a mix of both your suggested reviewers and other people who the Program Officer selected. In the end, many factors determine who will be the ad hoc reviewers for a specific proposal: on the one hand, some reviewers you suggested may be unavailable or may be perceived to have a conflict of interest by the Program Officer; on the other hand, the Program Officer may be unable to get any of the people he or she wants and thus end up using most of the names you suggested.

4.2.3 Program Officers

NSF recruits active researchers from universities and other institutions to serve the role of Program Officers and by default, all NSF Program Officers are also experts in specific research domains, just as the panelists and the ad hoc reviewers. There are two types of NSF Program Officers: permanent and visiting. Permanent NSF officers have detailed knowledge about the history and the portfolio of a program, and have a vision of where the program is going in the long run. Visiting NSF Program Officers bring new perspectives and fresh ideas to the program, as they are usually at NSF for only 1–3 years, often on leave from an active teaching or research position. Although both types of Program Officers bring to the table their own research expertise and perspectives, they tend to take a more neutral view toward specific proposals with regard to research approaches or methodologies in making their funding recommendations.

While panelists' and ad hoc reviewers' main job is to provide expert opinions to the NSF, and to a lesser extent some useful comments to the investigators, it is the Program Officers who make a final recommendation on the funding of a proposal. Thus, the panelists and the reviewers are there to help the NSF Program Officer make an informed decision, but are not themselves the decision makers. It is also not their task to help the investigator improve the project, as discussed earlier. In some panels, the Program Officer may also explicitly tell the panelists not to comment on proposal budget, as budget issues are dealt with by the Program Officer only after a positive recommendation is reached for a proposal (see Chapter 1, Section 1.3.4, and Chapter 2, Section 2.4.8, for discussion of budget issues).

At NSF the Program Officer's job is to make funding recommendations, based on panelists' and ad hoc reviewers' comments. This is quite different from the process at NIH, where each proposal is associated with a "priority score" that is determined by the panelists at the panel meeting and the Program Officer has little flexibility in changing the status of a proposal given a fixed priority score. NSF Program Officers have more degrees of freedom in making funding recommendations, because NSF does not use a rigid scoring system. In a way the Program Officer can act like the editor of a journal in making a positive or negative decision. In most cases, the Program Officer would agree with the panel's comments and its categorization of a proposal (see Section 4.3.1 below). But in some cases, especially when there are conflicting or inconsistent comments from the panel or the reviewers, the Program Officer needs to take a hard look at all the reviews and the original proposal, and then make an informed judgment and recommendation. This is a stage at which the Program Officer is actively using his or her expert knowledge and research expertise.

Although NSF funding decisions rely on the opinions from the panelists, the ad hoc reviewers, and the Program Officers, not all programs at NSF use all three sources of information or use them equally. For example, some proposals may involve only panel reviews and no ad hoc reviews (as discussed above), and some proposals may involve only internal reviews by the Program Officers. Conference proposals may or may not need to be reviewed by a panel or ad hoc reviewers, depending on whether the Program Officer feels the proposal warrants such review and whether the proposed budget exceeds a certain limit (e.g., above $100k in total cost ; see also Chapter 1, Section 1.3.3.5). Exploratory projects (EAGER) and time-sensitive projects (RAPID) in most cases do not need to be reviewed by a panel, and several Program Officers can get together and discuss the merits of EAGER and RAPID proposals. If a proposal is sent to a panel for review, it requires a minimum of three reviews, regardless of whether they are from the panelists or the ad hoc reviewers.

4.3 PANEL MEETING, PANEL SUMMARY, AND CO-REVIEW

4.3.1 Panel Meeting

A few months after your proposal reaches NSF, the panelists get together (usually at NSF headquarters) to discuss the merit of your proposal (see Figure 1.2 in Chapter

1 for the timeline). At this time, the Program Officer should have gathered all necessary information associated with your proposal, including at least three sets of comments (e.g., from the two panelists assigned to your proposal plus one ad hoc reviewer). Note that only the panelists meet in person with the Program Officer at the panel and the ad hoc reviewers are not invited to the meeting. Occasionally a conference call may be placed at the panel meeting for a special panelist, who, for various reasons, cannot make it to the meeting in person or can only review a few specific proposals. NSF now has several methods of holding panels, including using online tools such as Skype, Second Life, and WebEx, as well as traditional videoconferencing and teleconferencing.

Normally, you may not care what exactly goes on in the actual review of your proposal at the meeting, but it might be helpful to know some details, because what goes on during the panel can determine what you should do when you revise and resubmit your proposal. As mentioned earlier, each proposal is assigned to two or more panelists, the exact number of which can vary quite a bit from panel to panel and from program to program. In most cases, there is a primary (or lead) reviewer, and there is a secondary reviewer (or even a tertiary reviewer). For some panels, the primary reviewer is also the so-called "scribe"—the panelist who writes a summary of the panel discussion (the Panel Summary), whereas for other panels, the secondary reviewer may be the scribe (in still other panels, the scribe may be a panelist who is not a reviewer of the proposal but who has read the proposal). For many panels, both the primary reviewer and the secondary reviewer are required to submit written reviews before the panel meeting just as the ad hoc reviewers do, but their reviews can be modified or revised as a result of panel discussion. For some panels, the secondary reviewer is not required but encouraged to provide a written review. During the panel meeting, the primary and secondary (and tertiary, if applicable) reviewers will all comment on the proposal and discuss its merits and impact. As you can see, there is quite a bit of flexibility at NSF with regard to how panelists are assembled and required to do their job.

At the beginning of the meeting, the Program Officer welcomes the panelists and lays out the ground rules of review and the role of the panelists, including conflict of interest issues. The panelists are asked to evaluate the proposal on the basis of the NSF review principles and criteria, that is, the three guiding principles, two review criteria, and five review elements for evaluating the proposed project (see above and Chapter 2), commenting on the substance of the proposal, including scientific significance and feasibility of the proposed project. If a panelist has a personal relationship with you (the PI, co-PI, or any senior personnel), has published with you in the last 4 years, or is from the same institution as you, the panelist is considered to have a conflict of interest and cannot review your proposal; he or she will step out of the meeting room when it comes to the discussion of your proposal. The primary panelist usually leads the discussion by giving an overview of the proposal and then describing the proposal's strengths and weaknesses in a few minutes. This panelist will be the one who explains your project to the rest of the panel and convinces them that your proposal is excellent (you have got an advocate) or poor (you have bad luck). The secondary or tertiary panelist then chimes in, agreeing or disagreeing with the primary panelist and adding further comments. One of them could also be asked to summarize what the ad hoc reviewers say with regard to each of the two main NSF criteria. Then the rest of the panel members will either show support or disagreement for the points made by the primary or second panelists and the ad hoc reviewers, or they may have specific questions or ask for clarifications. The entire discussion of a proposal usually lasts about 15–20 minutes, depending on how many proposals the panel is handling and how varied the reviewers' opinions are for a given proposal. In some cases, discussion of a single proposal could last for more than half an hour, while in other cases, it is so straightforward that no serious discussion is needed. The former case typically occurs when panelists have divergent opinions and become engaged in debating the pros and cons of a proposed project (in some rare cases, the panelists may not be able to resolve their differences, which may be reflected in the Panel Summary; see below). The latter case usually occurs with very poorly written proposals that have received consistently negative reviews before the panel, and these proposals are often "triaged" without detailed panel discussion.

Typically, after discussion of each proposal, the panel will place the proposal in one of the ranking categories (e.g., high priority or highly competitive, medium priority or competitive, low priority or non-competitive). Note that there is a lot of variability in terms of what these categories are called: some panels will assign different labels or simply assign numbers from 1 to 5. The initial ranking of a proposal may be revised by the panel when all proposed are discussed (see below), and the ranking should be consistent with the panel summary and the individual written comments by the panelists (hence the panelists may revise their written comments after panel discussion, although the ad hoc reviewers' written comments are left untouched).

In placing a proposal into a category, the panel is reminded by the Program Officer that both Intellectual Merit and Broader Impacts, the two NSF review criteria, should be considered equally important (see discussion in Chapter 2 and earlier in this chapter). To the extent that your proposal stands out in its broader impacts, your chances of success are increased compared with other proposals that are equally excellent on other dimensions (see also discussion in Chapter 2, Section 2.5.3). As mentioned earlier, NSF now leaves it to the PI to decide what specific activities qualify as broader impacts of the project. However, panelists and reviewers need to consider the societally relevant outcomes of a project in light of NSF's Strategic Plan (see http://www.nsf.gov/news/strategicplan/index.jsp). Such outcomes include increased participation of women, persons with disabilities, and underrepresented minorities in science, technology, engineering, and mathematics (STEM); improved STEM education at all levels; increased public scientific literacy and public engagement with science and technology; improved well-being of individuals in society; development of a globally competitive STEM workforce; increased partnerships between academia, industry, and others; increased national security; increased economic competitiveness of the United States; and enhanced infrastructure for research and education. Reviewers are reminded that these are illustrative examples and investigators may include other appropriate examples not covered by these.

The exact categories or number of categories may differ from panel to panel and from program to program, but there is no general vote or scoring involving all panelists, as there is at NIH panels. The categorization usually begins with the primary panelist suggesting a category, and the secondary reviewer agreeing or disagreeing, followed by the panel's concurrence. In cases of disagreement, the final ranking may be a compromise between different panelists' opinions (or in rare cases, a proposal may become unranked due to the fact that the panel could not agree). It is NSF's common practice that panels come back to revise their rankings, before the end of the panel meeting after all proposals are reviewed. This way the panel can compare the rankings against all proposals reviewed and see the big picture. This process often leads to changes in the ranking, mainly in moving the borderline cases around to one or another category. The Program Officer could also ask the panel to provide a finer rank order of the proposals within the highly competitive or competitive categories, so that recommendations at a later stage may become easier in light of constrained budget.

At the end of the meeting, the panelist who is the scribe will write a Panel Summary, which is then read and amended by all panelists involved with the proposal, and is approved by the Program Officer before the end of the meeting. Each Panel Summary may look slightly different from another one, given the different focuses of the discussion at the panel. In addition, different programs may have different Panel Summary templates that they provide to the panelists/scribes, and if your proposal is co-reviewed by two or more panels (see Section 4.3.3 below) you will see such differences.

The above procedure of discussion at the panel meeting applies to all types of proposals, including regular research proposals, conference proposals, collaborative research projects, and so on. NSF requires all proposals be evaluated by the same set of review principles and criteria, and as such, all types of proposals are discussed with regard to intellectual merit and broader impacts. However, different types of proposals will be treated slightly differently from one another, in that panelists understand the different focuses and outcomes of different types of proposals. For example, for conference or workshop proposals, the research itself is relatively short-term, and the impact mainly lies in its integration of traditionally different disciplines. Reviewers and panelists ask whether the occurrence of such a conference or workshop can deliver new synergies across fields or subfields, and whether the conference organizer has assembled the right people and expertise for cross-talks and conceptual cross-fertilization. For collaborative research projects, reviewers take a close look at whether the two or more investigators or institutions that are involved in the project really are the right match or the right combination for the type of work proposed, and whether this combination provides unique and complementary expertise, resources, and synergies for generating promised findings and outcomes from the collaborative project. They will question whether the same project can be carried out in one institution rather than by multiple institutions, and what added values the collaboration can bring to the research. The point of "value addedness" is important for all types of collaborative projects, especially for transnational collaborations (see discussion in Chapter 7, Section 7.3.2).

Although you have no control over the actual discussions at the panel meeting, you should be able to see from the Panel Summary how your proposal was treated by the panel, and how and whether it was discussed at all. Because of the flexibility associated with how NSF panels are run, it is important for you as the investigator to know that there may not be a single procedure that applies to the review of all proposals, varying from the number of reviewers assigned to each proposal to the ranking categories used and to the funds available to each program.

4.3.2 Panel Summary

The Panel Summary provides a succinct description of what the panel deliberations are with regard to the intellectual merit and broader impacts of a given proposal, laying out both the strengths and weaknesses of the proposal. The Panel Summary also gives the ranking or category that your proposal belongs to. Panel Summaries are often accompanied by a Context Statement (or Program Officer Notes), which provides you with information such as when the panel meeting took place, how many proposals were reviewed, how many of them were placed into each ranking category, what categories were used, and what each category meant. If you have questions regarding any of this information, you should check with the Program Officer. This information gives you a good idea of what your proposal was up against and how competitive the process was, which in many cases will make you feel better (you are not alone in getting the bad news!). In this economic climate and the low funding rate today, most of us get negative news from NSF, especially the first time around, so negative outcome is the norm rather than the exception.

You should start by reading the Panel Summary first, and use it as a guide for you to read through the individual reviews. A good Panel Summary should give you the context of the discussion, and shows what the panelists were thinking when your proposal was on the table for discussion. The panel may not always agree with the ad hoc reviewers, and if you read the Panel Summary carefully along with the individual reviews you will see how much attention and weight each review received during the panel discussion.

In some cases, you may be disappointed to see that no detailed summary of the panel discussion is available, due to the fact that your proposal was not discussed at the panel at length. NSF panels do not routinely triage half of the proposals as NIH does, but occasionally they do so to some proposals that have received consistently negative reviews with very low ratings. In those cases, instead of a detailed summary the Panel Summary may contain a clause like the following: "The panel agreed with the substance of the written comments made by the reviewers and therefore declined to further discuss the proposal at the meeting." Such a Panel Summary gives a clear signal that you should not try to resubmit your proposal without significant and major changes to the methodology, hypotheses, or theoretical framework of your proposed research.

Having read the Panel Summary and the reviews, you now have the task of thinking hard about if and how you should revise the proposal in order for it to be on a positive trajectory next time around (see Chapter 5 for further discussion of the revision process). If the reviewers' comments are consistent, you will have an easier job to do. If the reviewers do not agree with one another, you need to determine, as an investigator, which reviews contain more important or more critical points of view that should guide your next revision. If you can tell where the panel disagreed with the ad hoc reviewers, this is also helpful, because the panelists usually hold more weight in the review process than the ad hoc reviewers; they are the ones present at the panel discussion whereas the ad hoc reviewers are not. We can imagine several scenarios at the panel: (1) The panelist is positive about your proposal and the ad hoc reviewers are also positive, so the panelist plays the role of an advocate for your research, and you are in luck. The more enthusiastic the primary or second panelist is about your proposal, the stronger an advocate you have. (2) The panelist is negative about your proposal and the ad hoc reviewers are also negative, so this is bad news for you. (3) The panelist is positive but the ad hoc reviewers are negative, so this will depend on how enthusiastic the panelist is about your proposal, but overall your chances will be reduced by the negative ad hoc reviews. (4) The panelist is negative and the ad hoc reviewers are positive, so here again your chances are not so good, even if the positive ad hoc reviews can counteract the overall ranking a bit (of course, the panelists themselves may also disagree with each other). In other words, for your proposal to be successful, you need to have a strong voice from one or two panelists, and these voices must also be echoed clearly in the ad hoc reviews. A lukewarm review of your proposal by one or two panelists almost always leads to the demise of a proposal. The final ranking of your proposal is thus the composite of the panelists' opinions, the ad hoc reviews, and the interactive discussions at the panel.

4.3.3 Co-Reviews

Sometimes you don't get just one Panel Summary, but two or more. This may be because at the time of submission you had selected a secondary or tertiary program, in addition to the primary program, to review your proposal. If you had not made such a request and you still get two or more panel summaries, it is because the Program Officer has decided that your proposal is appropriate for co-review by one or more other programs (Program Officers often make such a decision without informing the investigators). In some cases, you may even find out that another program has taken over your proposal as the primary program to evaluate your proposal and the program you submitted to has become the secondary, because the original Program Officer determines that your proposal fits the mission and scope of the other program better (and this is agreed upon by the other Program Officer). In any case, all of this indicates that your proposal cuts across traditional disciplinary boundaries (as designated by the NSF programs), and this is a good thing.

Each of the above situations requires a co-review of your proposal. This means your proposal will be reviewed by two or more programs/panels. Typically, the primary program administers both the review and the funding process just as when no other programs are involved, while the secondary and tertiary programs simply do a co-review. In other words, it is the responsibility of the primary program to send your proposal out for ad hoc reviewing, coordinate the panel, and discuss the budget and management issues of your project in case your project is funded. What the co-review entails is that your proposal will be reviewed not just by one panel, but by two or more panels, each of which follows the steps discussed in Section 4.3.1. Whether your proposal will be first discussed by the primary, secondary, or tertiary program depends entirely on when the panel meeting of each program takes place, independent of the specific proposals under review. The co-review will involve the relevant two or more Program Officers sitting in at each other's panel meetings when your proposal is discussed, so that they know how your proposal fares at the panel of the other program or programs. Each Program Officer is also free to invite additional ad hoc reviewers to comment on your proposal, in addition to the panelists assigned to your proposal from each program. That's why sometimes a proposal ends up with 6–9 reviews rather than 3 reviews.

When all the relevant panels have met, the Program Officers from the primary, secondary, or tertiary programs of your proposal will sit together and discuss the co-review results, and decide whether they would like to co-fund your project (co-review is mostly associated with co-funding; see below). Again, imagine several situations under which your proposal is reviewed by two programs: (1) Both the primary and second programs like your proposal, so you are in luck. (2) The primary program likes your proposal, but the secondary program does not, so no co-funding is possible as the secondary program does not want to kick in support; in this case, the primary program can either ignore the negative comments from the secondary program and fund your project alone (less likely), or reject your proposal based on considerations of the negative comments (more likely). (3) The secondary program likes your proposal, but the primary program does not, so again no funding is possible. In this case, the secondary program can ignore the negative comments from the primary program and indeed take over as the primary program of your proposal and fund it (very unlikely), or have the primary program reject your proposal (most likely). The likelihood of one way or another in the second and third situations will also depend on how enthusiastic one of the co-reviewing programs is about your project. Sometimes a program may have clear reasons to believe that the other panel or panels are biased toward your proposal, and thus will make a strong case for your proposal.

When should you request a co-review of your proposal? As mentioned, sometimes this is determined solely by the Program Officer. But if you have a choice at the beginning (deciding whether to select secondary and tertiary programs or not), you have to think through about the advantages and disadvantages of co-reviews. The advantages, as discussed, are that co-reviews of your proposal indicate the interdisciplinary nature of your proposal, and the possibility of co-funding by two or more programs. NSF likes interdisciplinary research and encourages co-review and co-funding. Co-funding also splits the financial burden of a program, especially in supporting large-scale projects. In addition, as discussed in the third situation, co-review allows for the possibility that other programs may like your proposal even if the primary program does not. This is likely especially when one panel lacks the expertise to evaluate your proposal whereas the other panel fills this gap perfectly. In practice, this situation is most beneficial when the primary program is only lukewarm about your proposal but the secondary program is highly enthusiastic about it. On the other hand, the disadvantages of co-reviews are just too obvious. The more panelists and reviewers you have for your proposal, the more likely your proposal gets picked at and criticized, which will definitely reduce your chances of success (see the above discussed situations). In today's funding situation, the proposals that get funded may often be the ones that have been ranked by all reviewers as excellent or close to excellent. Thus, co-review would work best only if you have confidence in the highly multi-disciplinary nature of your proposal and in the ability of your proposal to target the interests of multiple panels.

4.4 REVIEW DECISIONS

Once the panel meeting is concluded, the Program Officer's real job starts. He or she needs to do a thorough analysis of all the reviews as well as the Panel Summary associated with a proposal, and writes a Review Analysis for each proposal. The Review Analysis is an internal document that the Program Officer shares with other NSF colleagues, mainly the Division Director or Deputy Division Director. It contains general information about the review panel, statistics of submitted proposals, panel rankings (information in the Context Statement; see Section 4.3.2), and specific information about the strengths and weaknesses of a proposal with regard to the NSF review principles and criteria, along with a final recommendation for or against funding of the proposal. As mentioned, if the comments from the panel and the ad hoc reviewers are consistent, the Program Officer has a somewhat easier job. If the comments are inconsistent or conflicting, the Program Officer needs to examine the various arguments in detail, weigh in with his or her own opinions, and then make an informed decision. The Review Analysis forms the basis for the Program Officer's recommendation for funding.

For some large interdisciplinary panels (e.g. CRCNS, NSF 11-505) that involve multiple programs, there may be several Program Officers who will gather together for discussion before the Review Analysis is written for a proposal. These discussions often take into consideration not only the ad hoc reviews and panelists' comments but also multiple Program Officers' opinions. Thus, a proposal that is reviewed well at the panel may not necessarily get funded because of the different perspectives that one or several Program Officers take toward the proposal.

Although officially the Program Officer only makes a recommendation for funding, with which the Division Director needs to concur (or not concur, rarely), the Program Officer's recommendation normally determines the status of a proposal. The opinions from the panel and the ad hoc reviewers, and the panel's ranking of the proposal, are all taken into consideration in the Program Officer's Review Analysis, as discussed above. However, in doing the final recommendations, the Program Officer may need to re-rank some proposals, considering other constraints including priorities of the foundation, portfolio of the program, past funding history, and availability of budget. Information regarding these additional constraints is not made available to the panel or the ad hoc reviewers, and must be dealt with post-panel by the Program Officer. The funding priorities of the NSF may change slightly from year to year, and that depends a lot on the available budget, and sometimes even on mandate from the Congress. From the Program Officer's (and NSF's) perspective, it is important to build a balanced portfolio in terms of where the money goes (types of institutions and locations), who receives the money (type of investigators, junior vs. senior, women or underrepresented groups), and what type of cutting-edge research is supported with the limited budget. Additional considerations could also include the judgment by the Program Officer on a proposal's potential for transformative advances in a field and its capacity for building new and promising research areas or research infrastructure, along with further considerations of the review principles and criteria. Because NSF does not rely on a strict scoring system (e.g., the priority scores of NIH), the re-ranking of proposals based on these additional constraints is possible, although such re-rankings must be justified on grounds of both the scientific quality of the proposal and other practical considerations.

One very practical consideration is availability of budget. Depending on the available budget for a given year or a given cycle, the Program Officer will move the funding bar up or down to a specific percentage point. The actual rate of funding at NSF varies from 10% to 20% for most programs, although this number differs from year to year, and in the past few years the NSF's funding rate has not been too good (except in the spring of 2009 when there was the economic stimulus money from the government). The funding rate determines if some very good, fundable proposals will in fact be funded. One of the most frequent sentences in the Program Officer's Review Analysis, unfortunately, is "Given the above considerations, along with the limited budget available to the program, I do not recommend the proposal for funding." The budget factor also makes it unlikely for a program to fund multiple projects that have identical or very similar scopes, multiple projects from the same institution at the same time, or multiple projects from the same investigator or same group of investigators. This is because the Program Officer is required by NSF to build a balanced portfolio, with considerations of the priorities of the foundation and of the diversity of funding. Given the limited budget, the NSF Program Officer prefers not to have all the funds go to the same researchers, same institution or same type of institutions, a concentrated geographic location, or overrepresented groups of investigators. These considerations with regard to the portfolio of a program are important for investigators to understand. Their implications could include, for example, not submitting multiple proposals to the same program at the same time, not competing with colleagues from the same institution who may have similar research proposals, and collaborating with colleagues at other institutions, especially different types of institutions (e.g., smaller private schools with big public universities).

Chapter 5

Revising Your Proposal

5.1 WHY REVISION?

It is very frustrating to all who receive the bad news from NSF that their submitted proposal is not favorably reviewed and is declined for funding. "Months of hard work are wasted," you will say! It could be even more depressing when you hear the bad news for a second or third time on the same proposal. And you are not alone. As we mentioned in previous chapters, getting a proposal rejected today is the norm rather than the exception. However, there are two key points to success at NSF (or at any funding agency): first, don't give up, and second, know how to improve your proposal in a revision. For the first point, we know many investigators keep revising their proposal many times, and in this process, the research plan becomes clearer and better, which may lead to better research, which in turn increases the chances of success at NSF (so your months of hard work are not wasted even if you do not receive funding initially). The goal of this chapter is related to the second point, that is, to discuss strategies and tips that will help improve the proposal after one or several revisions.

Nothing replaces good ideas in science. But having good ideas without good exposition is not likely to lead to success at NSF, either. You need to know how best to communicate your ideas to the reviewers, panelists, and the program directors; the previous chapters have already touched upon many of the effective ways in which you can achieve successful communication. If you really have great ideas, and your NSF proposal is not in the top 10%, then most likely it is because your proposal is poorly written, the reviewers do not understand it, the program director does not find transformative science in it, or a combination of some or all of these. While we cannot help you make your scientific ideas better in this book, our goal in this chapter is to make your proposal stand out in style and in effective communication.

Having Success with NSF: A Practical Guide, First Edition. Ping Li and Karen Marrongelle.
© 2013 Wiley-Blackwell. Published 2013 by John Wiley & Sons, Inc.

5.2 FIRST STEPS IN REVISION

5.2.1 Reading the Panel Summaries and Reviews

In Chapter 4 we discussed what you should do when you receive the reviews and the panel summaries for your proposal from NSF. These materials give you a good sense of how competitive your proposal was and how the panelists viewed your proposal in the context of many other proposals. It is imperative that you study the panel summaries and the reviewers' comments in detail (some of us may prefer to put away the "painful" comments for a few days before reading them in detail). The panel summary should also contain information about how the panelists weighed the comments of different reviewers, especially in cases when the reviewers' comments were not consistent.

Once you read these materials, you have to develop a strategy to revise your proposal such that the negative reviews will become positive the next time around. Your goal here is not too different from revising a journal article to satisfy the reviewers' comments. One thing that you should keep in mind is that while in most cases a journal editor will attempt to track down an original reviewer for a second look at your revised manuscript, the NSF Program Officer cannot guarantee that your revised proposal will get a second review by the same original reviewers. This is because first, all NSF proposals are technically new proposals (unlike at NIH) and will be treated the same at each review cycle, and second, the NSF panels are dynamic and there is likely a quick reviewer turnover (see discussion in Chapter 4, Section 4.2.1, about how panelists are selected). Of course, these conditions do not mean that the NSF Program Officers make no attempt to find the original reviewers of your proposal (they do!). In NSF divisions where multiple Program Officers administer a single program, you are not guaranteed to have your revised proposal handled by the same Program Officer. In addition, because some Program Officers are at NSF for only a short time as rotators, it is also likely the case that a second, third, or fourth revision of your proposal will not be handled by the same Program Officer! Thus, as an investigator you should always, without knowing who will be on the panel or who may review your proposal the next time around, consider seriously whether you can address the reviewers' concerns and the weaknesses of your proposal as they are raised in the first round of review. The Program Officers generally have a good idea of the history of a proposal and what issues have been raised and what have or have not been addressed in the revision. So it is often a good idea to email or talk with the cognizant Program Officer about (1) whether it makes sense for you to revise and resubmit and, if so, (2) what to focus on in the revision.

If you have determined from the panel summaries and reviews that the overall ranking or tone is extremely negative, it would make little sense to send the same proposal back to the same NSF program, even with revision. It might be the case that you have sent your proposal to the wrong program in the first place, because your research simply does not address relevant issues of the program (see Chapter 1, Section 1.2.3, for selecting the correct program). In such situations, you may be better off considering other programs or other funding agencies, or radically restructuring your ideas for a new proposal.

5.2.2 Knowing What's Wrong and Shaping Your Message in a Revision

Once you have carefully studied the comments from the reviewers and the panelists, you should be able to decide whether it is the Intellectual Merit or Broader Impacts (two main NSF review criteria; see Chapters 2 and 4) that need to be reworked. With respect to Intellectual Merit, you should determine if it is the theoretical framework, the design of the study, the analytical tools to be used, the potential for transformative concepts, or the hypotheses and predictions that are in need of revision. With respect to Broader Impacts, you should determine if it is the lack of integration of science and education; the lack of broadening of participation of underrepresented groups in science; or lack of societal relevance that had raised reviewers' concerns. If you can see clearly that your proposal suffers from one or more of these aspects, then you can focus on revising the proposal on the specific questions. This type of careful analysis is also important because it allows you to see how much work will be involved in the revision: it is much harder, for example, to take a radically different theoretical framework than to redesign the experiments or to use a different statistical analysis. However, if you cannot clearly identify the major problems, or if the reviewers' comments are all over the place, then your revision job is much harder and chances of success the next time around are also unclear.

A very typical set of critical comments from NSF reviewers and panelists may include some of the following: (1) Lack of major contribution: your proposed work represents only baby steps in one direction, the scope is too narrow, and does not make a significant contribution. (2) Lack of clear theoretical framework: the reviewer is unclear what theoretical stance or position you are taking, or what *theoretical* contributions your study will make. (3) Lack of clear focus: your work is too ambitious, with aims scattered all over the place. (4) Lack of a good literature review or contact with the literature: the reviewer sees poor review of the extant literature (sometimes lack of reference to the reviewer's own relevant work!), or failure to specify how the proposed study is related to the literature. (5) Lack of right expertise in theory or methodology: your own research strength and background do not indicate you can carry out the proposed research (see also Section 5.3.5). (6) Lack of details: the reviewer cannot understand your model or experiment because you have given very vague descriptions. With regard to the last point, you should note that reviewers are not asking you to provide all the details (e.g., how long your experiment lasts) as you would in a peer-reviewed journal article (so that the study can be replicated by others); the reviewer simply wants details at a level that can facilitate his or her understanding of the meaning of your proposed study and its contribution to the existing literature (see further suggestions and tips in Section 5.3, below, on writing clearly).

Investigators who are familiar with the NIH model often ask the following questions: Should I indicate that my proposal is a revised submission? And if yes, how do I do so given that NSF does not allow a cover letter to indicate the revisions made? As we discussed earlier, NSF treats every proposal technically as a new submission, so tracking whether or not you have made changes to the previous proposal is not part of the review process. Indeed, sometimes the panelists may be told explicitly that they should focus on the quality of the current proposal irrespective of the proposal's history. This is quite different from academic journals' peer review process. However, from a practical point of view, it is almost always the case that first, the Program Officer knows the history of a proposal in its various submissions unless there was a change of Program Officers and the transition was abrupt, and second, the reviewers often would like to know what improvements have been made to the previous version of the proposal, especially if the reviewer has read the previous version (note that you might get new reviewers or new panelists, given the unavailability of previous reviewers or panelists; see Section 5.2.1). Even if you get a new reviewer for your proposal the second time around, the reviewer might be curious about whether the current version is any better than the previously submitted version, and whether efforts have been spent on working out critical issues that had been previously raised. For these practical reasons, it is therefore always a good idea to indicate that your second or third submission is a revised version! You should be aware that Program Officers may not be able to tell curious reviewers if your proposal has been previously submitted (because every proposal at NSF is considered a new proposal). Therefore, if you want to communicate to the reviewers that you have taken previous comments seriously, it will be up to you to communicate this to the reviewers directly in your proposal.

Now, with regard to how to indicate the revisions you made, we are brought back to the analyses that you have conducted after reading the reviewers' comments and panel summaries (see Section 5.2.1). If you are able to identify the major concerns raised in the reviews and by the panel, in the new submission, you can but are not required to highlight the changes that you have made to the relevant sections. No separate letters or pages are given to you for indicating these changes, but in the relevant parts of the proposal text (i.e., within the 15 pages!), you can add comments such as: "In response to previous reviews, we are now proposing to do X, Y, Z," where the X, Y, Z could be the methods considered, analyses used, or outreach efforts planned. Again, you can but are not required to make some general remarks, very briefly (in a few sentences), at the beginning of the proposal to the effect that your proposal is a revised version taking into consideration the constructive comments from previous reviewers, listing the approaches you have taken to tackle the issues or concerns raised by the previous reviewers or panelists.

5.2.3 Should I Submit My Revised Proposal Right After It Is Rejected? Which Cycle Should My Proposal Go To?

As we mentioned in Chapter 1, Section 1.3.5, the lifespan of a proposal going through submission, review, and decision is long, and as such, you might be saying to yourself that you'd like to revise and resubmit a proposal right after it is rejected.

In other words, you would like to resubmit your revised proposal to the immediate next review cycle. But before you decide to do so, you need to ask yourself two questions: (1) Are the comments from NSF critical and serious, or are they relatively minor? (2) Do I have enough time to work on the proposal between now (when you receive the comments) and the deadline or target date for the next NSF review cycle? (See Chapter 3, Section 3.1, for deadline vs. target date.) Obviously, these two questions are related, and after you carefully study the reviews and panel summaries, you should have a good sense of how much time you need to revamp the proposal.

In general, it is advisable to resubmit to the immediate next cycle only if you have minor comments that point to issues relatively easy to address. A significant advantage for doing so is that you are more likely to get the same reviewers or panelists to look at your proposal the next time around this way than if you wait for another 6 months. However, if the reviewers and panelists have raised significant concerns regarding the methodologies or theories of your project (e.g., you need to adopt a new approach or perspective, need to involve specific expertise; see Section 5.2.2), it is better for you to sit on it for a little while and wait for the next review cycle (e.g., you may need time to get another expert to join you for a collaborative research project; see Section 5.3.5 and Chapter 2, Section 2.7). In programs that operate on an annual cycle (that is, proposals are accepted only once per year), you should have enough time to revise and resubmit your proposal, whether the reviewer comments are major or minor. In any case, if you get encouraging comments from the panel and reviewers, you should check with the Program Officer whether it makes sense for you to revise and resubmit immediately.

At NSF, many regular programs have two review cycles, the spring cycle and the fall cycle, and the specific deadlines or target dates vary from program to program. If your proposal is reviewed in the spring, and you want to resubmit it to the fall panel, in most cases this could work because of the free time you'll have during the summer. But if your proposal is reviewed in the fall, and you would like to resubmit it to the spring panel, the time might be tight. In such cases you may need to ask the Program Officer before the winter holiday begins if you have not heard about the status of your proposal. One interesting thing about the spring versus the fall cycle is that some programs may receive fewer numbers of proposals in the spring cycle than in the fall cycle (perhaps fewer people want to work on proposals during the winter break, or more people want to work during the summer break). This could mean that the competition in the spring cycle is slightly less fierce than in the fall cycle. An interesting fact to note is that the NSF's fiscal year ends on September 30 of each year (the new fiscal year begins on October 1, in line with the U.S. government's fiscal calendar), and the proposals reviewed in the spring cycle could get an extra boost if there are emergent funds that are not originally budgeted for a program; the NSF internal "close-out" date for the fiscal year is typically mid-July of each year, so that extra funds can be processed before September 30.

Some NSF programs may also have mechanisms other than time of the year for determining review cycles; for example, in the Directorate for Computer and Information Science and Engineering (CISE), many programs adopt three proposal submission windows based on the size of the project: Medium Projects (to be submitted during the second half of September each year), followed by Large Projects (submitted during most of November), and Small Projects (submitted during the first half of December). Therefore, even if you want to revise and resubmit immediately, you have to wait for a few months for the next review cycle to begin, unless you change the size of your project. On the other hand, this gives most people sufficient time to work on revisions, even if the reviewers' comments are substantial.

5.3 STRATEGIES AND TIPS IN WRITING A REVISION

5.3.1 Focus on the Major Issues

As mentioned in Section 5.2.2, we suggest investigators examine the panel summaries and reviews to determine the major issues raised by the reviewers and develop a focused plan to address these issues. In a revision you may not (and sometimes cannot) address all the issues raised especially if you have inconsistent comments from reviewers, but you do want to make sure to cover the major issues such that the revised discussions of the theoretical framework, the design of the study, the analytical tools to be used, or the hypotheses and predictions, will be positively perceived by the reviewers in the new round of review. Revised proposals have the best chance of success if the previous comments converged on a specific problem that is addressable with some efforts. Although every proposal is different with regard to what problems there may be, there are some general strategies and tips that can be useful to the revising of your proposal. We discuss a few of these below. Some of these tips may also be useful when you first write a proposal rather than when you start to revise one.

5.3.2 "A Picture Is Worth a Thousand Words"

You, as an investigator, may sometimes complain that the reviewers did not understand your proposal's great ideas, misunderstood the methodology, or failed to see the significance or implications of the study. While this may well be true, you should still think that the fault is yours, for two reasons: (1) the reviewer, especially the panelist, may be someone who is not in your specific research field (see discussion in Chapter 4, Section 4.2.1, on the selection of panelists) and (2) you simply did not make your ideas, methodology, or significance clear enough for non-specialists. Now, most reviewers are very conscientious, but they may also be under time pressure when they read your proposal (most experts are busy individuals!) and so may have missed some points in your proposal that are obscure or are not made crystal clear. This is especially likely to occur when the panelist needs to read a large amount of proposals within a few days (see discussion in Chapter 4). So the key is to write with clarity, not only for the expert in your specialized field but also for the "generalist" in your broader field, and convey, in simple and plain language, what the project is about, what theoretical and societally relevant significance it has, what exactly you plan to do, and how you can do it.

We say "a picture is worth a thousand words," and this applies to complex scientific proposals as well. If the reviewer complains about your theoretical framework being underspecified or unclear, you may want to provide an illustrative diagram that showcases your theory or hypothesis. If the reviewer takes issue with the design of your study, you may want to put in a flowchart illustrating the material, procedure, and protocols to be used so that no misunderstanding will result. Finally, if the reviewer has an issue with the feasibility of your research and the time line within which your work can be effectively conducted, supply a table or a flowchart having specific tasks aligned with specific times/months of the year. Other domains where diagrams and illustrations are particularly useful would be overall project plan and project management plans (especially for large-scale projects that involve multiple PIs and multiple research sites). The types of illustrations include traditional figures (bar charts, histograms, plots) for reporting results, Venn diagrams, flowcharts, sketches, and 2D or 3D images. Make sure that you explain the illustrations with some amount of text, but not too much (i.e., your pictures and diagrams should be mostly self-explanatory). If you have space, it does not hurt to include a figure or picture of the equipment that you use, especially if the equipment is rare or of a

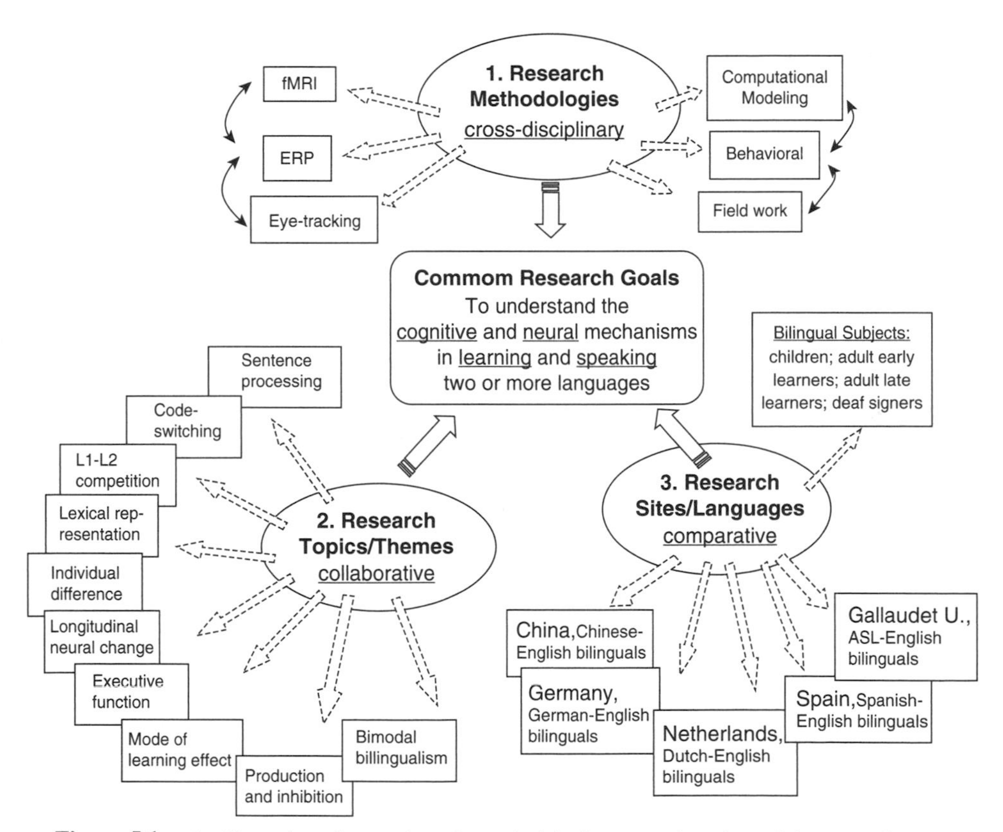

Figure 5.1. An illustration of research goals, methodologies, research topics and themes, and collaborating sites.

Task*	Year 1		Year 2		Year 3	
	Jan.–June	July.–Dec.	Jan.–June	July.–Dec.	Jan.–June	July.–Dec.
A	×					
B	×	×				
C		×	×			
D			×	×		
E				×	×	
F				×	×	×
G					×	×
H						×

Figure 5.2. Example showing how to present a project's timeline in a proposal (tasks and dates are for illustration purposes). *A, literature review + theoretical integration; B, Experiment One material, design, testing; C, participants testing + annual report to NSF; D, data analysis + conference participation; E, Experiment One writeup + Experiment Two preparation; F, new participants testing + annual report to NSF; G, new data analysis + conference participation; H, Experiment Two writeup and final report to NSF.

new technology. You do not need a figure on every page, but one illustration every three pages (e.g., about five for a 15-page NSF proposal) would just be fine. Figure 5.1 provides one such example from a previous successful NSF proposal submitted by one of the authors with colleagues.

This figure shows the Common Ground Model that guides the research project, along with collaborative and integrative research themes, cross-disciplinary and complementary methodologies, and comparative and unique languages/subject populations, all in service of achieving the common research goals (solid arrows indicate connectedness, connections within 2 and 3 and between 1, 2, and 3 are not marked for visual clarity). Such a graph allows the reviewer to keep in mind the complex project with some organizing principles, and to relate the text to this graph any time during reading of the proposal. A figure like this not only makes the whole text very clear, but also helps to save verbal descriptions.

Figure 5.2 presents another quick example of how you can illustrate a project's timeline for a typical 3-year NSF project. Without any complicated graphic design like Figure 5.1, this figure in tabular format shows clearly the task by time of year, and the planned execution of each task by the date. Note that the tasks listed here are for illustration purposes only and the details will depend on specific projects. You may also have a more fine-grained breakdown of your tasks and of your timeline in each year.

5.3.3 "Do Not Make Your Contribution More Informative Than Is Required"

Many investigators, especially early-career scholars, are excited about their ideas of research and when they write a research proposal, they attempt to provide a lot of information in the text, hoping that the reviewers will become equally excited. There are two potential issues that need to be considered in this context. First, the panelists are faced not with one single proposal, but 12–20 other proposals, many of which also contain exciting ideas. The key will be to not provide more information, but to provide better information in a concise and clear fashion. Thus, clarity and coherence in the expression of ideas are much more important than the sheer amount of information expressed (the discussion in Section 5.3.2 on having illustrations and figures shows one such means to achieve clarity). Second, since the page limit (15 pages for NSF standard proposals) is strictly enforced, you need to carefully think about what is most important to write about and what can be left out. The critical theoretical framework and hypothesis must be provided, some novel technology or methodology needs to be discussed, a discussion of the literature must be provided, and the broader impacts of the study should be made clear, but other details may be left out depending on their relevance to the central issues (e.g., how many participants will be tested in a study, where they may come from, how old they are, and what specific devices will be used for the experiment, all of which could be important but not essential). You need to achieve a good balance about giving the big picture and providing the fine details, and this often requires you to put yourself in the position of a reviewer, to think what will make your ideas stand out against all other proposals.

The Berkeley philosopher Paul Grice advocated the cooperativeness principle for conversation, and one of Grice's Maxims is the Maxim of Quantity: make your contribution as informative as is required, but not more informative than is required. This maxim should be observed by all grant writers! In many cases, more is not better, but less is more. Investigators who attempt to squeeze more information into the proposal by using smaller fonts, narrower margins, narrower spacing, website links, or any other means to increase the amount of information, will not only raise eyebrows at NSF (proposals may be returned without review after compliance check; see Chapter 4, Section 4.1), but also give reviewers a headache and make your proposal less appealing.

One final remark about being cogent with your argument: Be very careful in making categorical statements such as "I'm the first to develop this approach," "this work is the first of its kind," or "this approach is entirely new, ground-breaking and revolutionary." It is a very common comment from panelists and reviewers that an investigator has not paid attention to literature that is highly relevant to the project, and chances are that there might be previous studies that would make your categorical statement invalid. So rather than making bold statements like these, you should let the reviewers and panelists come to the conclusion that your proposed work is truly innovative, novel, and revolutionary (if it is); your job instead should focus on having a good and balanced literature review, comparing your approach to previous approaches and highlighting its uniqueness and contributions.

5.3.4 Have a Clear Structure in Your Proposal and Have a Clear, Informative Title

Providing a balanced discussion of theory and experimentation at the right level is essential for the reviewers to get the big picture while appreciating some details, as discussed above. Technically, this often requires the grant writer to structure the discussion clearly so that it will enable the reviewers, especially the panelists, to read the proposal easily. Clarity is achieved only when the reader can feel the natural flow of text, with regard to both the overall framework and the accompanying details. It is important to have properly structured section headings, subheadings, and a good numbering system for outlining theories, hypotheses, experimental procedures, implications, and impacts (note it is now mandatory for proposals to include a separate section of discussion of Broader Impacts, so you also need a separate section heading "Broader Impacts of Proposed Activities"). Remember to use simple words, simple sentences, and plain language to describe your complex ideas (along with the use of figures and illustrations, discussed in Section 5.3.2). Reviewers don't like long paragraphs, convoluted sentences without clear focus, and text running through the pages without headings or subheadings. And always remember to tie your specific procedure and experimentation to the theoretical questions asked, toward the end. Finally, make sure that your proposal is free of typographic errors or grammar errors. (Run a spelling check! Typos indicate poor preparation.) The clearer your proposal is, the faster the reviewer can read and understand it, and the better the reviewer's impression will be. Think of your proposal as the one at the end of the stack sitting on the panelist's desk (he or she just wants to get through!). The last thing you want a panelist in this situation to do is to read the same text or paragraph three times in order to figure out what you might mean.

It cannot be emphasized enough that you need to know how to communicate your ideas to researchers who might be slightly outside your field. In Chapter 4, Section 4.2.1, we discussed how panelists are selected and the fact that not everyone on the panel is an expert in your specific research field. This is very important to know, as it means you are not writing a technical journal article for your close colleagues, but rather communicating an exciting idea to the NSF and asking for funding for that idea. Clarity, in simple scientific terms (avoid too much jargon and too many acronyms!), with a big picture and some details, will be the ingredients that form the basis of good understanding. Once you have that understanding built up, the reviewer or panelist may be able to judge the merit of your proposal even if he or she is not directly in the subfield of your research.

Some proposals suffer from the lack of a good title, and this may not seem like a big deal to some researchers. However, having a good and informative title may be more important than you think. For example, when the Program Officer selects reviewers and panelists for a given proposal, sometimes the key factor that determines who the reviewer is may be the title, given that it is impractical at the initial stage for the reviewer or the Program Officer to look into the details of a proposal. When the panelist receives the spreadsheet of proposals from NSF and tries to determine whether a given proposal fits his or her own expertise, the panelist scans through the proposal titles and makes an initial decision. Thus, if your title is not informative or is misleading, you may end up getting a panelist who does not really have the right expertise to evaluate your proposal (though be mindful that panelists are not always experts in the specific domain of your proposed research, as discussed). Another important aspect about titles is that when your proposal is funded,

NSF posts your project title and abstract on the internet, searchable by the public, and titles play an important role in determining the search results.

5.3.5 Changing to a Collaborative Project for the Revision?

A very frequent comment from NSF review panels is that a proposed project lacks the right expertise in either theory or methodology; in other words, the PI really does not have the qualifications to carry out the research alone. The recommendation may be that the investigator should find an expert in the relevant domain. This would be a perfect time for you to seek a collaborator or collaborators who can fill a gap in the knowledge and skill required to execute the project. Whenever this type of comment appears in the reviews, you should make every effort to find the right people with the right set of research background and skills. There are several ways you can go about doing this.

First, you can identify a colleague from another institution who is willing to do a Collaborative Research project with you. NSF allows two or more PIs from different institutions (multi-site PIs) to contribute to the common goals of a project (in addition to Collaborative Research projects, you can also involve a colleague from another institution through the subaward or subcontract mechanism; see Chapter 2, Section 2.7). Second, if you have a colleague who has the right expertise within your own institution, you can involve him or her as a co-PI, bearing in mind that the co-PI must be able to fill an expertise gap. Finally, you can try to find a junior colleague who might have recently come out of graduate school to be a potential postdoctoral fellow to work with you on the project. The potential candidate should possess just the right skills required to fill the void as pointed out by the reviewers. This person should understand that once your project is funded, he or she will come to work with you to contribute specifically to some aspects of the research. Clearly, the above three models (involving collaborators from another institution, from within the same institution, or from a potential postdoc) place different demands on the PI's role, involve different levels of expertise, and require different degrees of collaboration and oversight. Only you can determine which model is the right one, given the comments provided by the review panel.

A word about Collaborative Research projects. These projects involve two or more institutions and NSF generally requires them to be larger in scale and in scope than regular projects (and there may be more funding for them, too). The key to deciding on having a multi-site collaborative research project should be that the two or more PIs have unique and complementary expertise that is well suited for the project, in either disciplinary knowledge or methodologies. Another consideration is whether the collaboration will bring new synergy to the project, such that new approaches might emerge, or new resources become accessible to the broader community as a result of the collaboration (e.g., high-quality databases or new analytic tools). Also make sure that the contribution of each collaborator is clearly specified, and that there is a good plan for efficient and effective coordination via meetings and other channels of communication (e.g., frequent communication via Skype, phone calls, conference gatherings). Finally, you shouldn't simply go for the name or reputation of a colleague and think that adding him or her to the project makes the project look good. If a big name does not have the necessary and complementary expertise required, the value added by having him or her on the project may actually become suspicious to reviewers and therefore decrease the overall chance of success. Such considerations are especially important to certain NSF programs where collaborative efforts are the prerequisites for proposal submission (e.g., Collaborative Research in Computational Neuroscience, CRCNS, NSF-11-505, or Cyber-Enabled Discovery and Innovation, CDI, NSF-11-502).

If you do submit a Collaborative Research project, remember that there are small things that you should pay attention to in writing the proposal, such as using "we" instead of "I," "the investigators" instead of "the investigator," and "our laboratories" instead of "my laboratory." This way the reviewers will know that you have worked together and are planning the joint proposal together, rather than pulling two separate research strands into one. Frequently, individual collaborators write separate pieces of the proposal and merge them together, without doing the necessary integration or even changing "I" to "we," which raises red flags to the panelists.

5.3.6 Have a Colleague Read Your Proposal

When you think you have successfully addressed the reviewers' comments and are ready to resubmit, it is a good idea to ask a close colleague or two to read your revised proposal (see also Chapter 1, Section 1.3.1). Your collaborator of course gives you valuable comments, perhaps already along the process of grant writing, as a co-PI or consultant. But someone who is slightly outside your research domain might in some cases be even more helpful, considering what has been mentioned again and again about the expertise and composition of the panelists. This individual may have general grantsmanship due to experience from elsewhere (e.g., having served as a PI or co-PI, or as a panelist for NSF or other foundations). The more similar such an individual is to a panelist, the better it is for him or her to give you critical and useful comments. In other words, if the individual works in an area related to your research but not directly in the domain, and if the individual knows well the grant writing and review processes, his or her comments will be most helpful. It is interesting to note, in this context, that some Asian universities have started to seek services of this type by requesting external reviewers *before* the proposal is submitted to agencies for formal review, and in some cases even before the complete proposal is finalized (e.g., one of the authors receives such requests frequently from universities in Hong Kong). Sometimes expert comments provided to even an outline of a proposal (with some details fleshed out) could guide and shape the development of a good proposal and consequently increase the success of the proposal.

Your colleague's comments, even if they are not directly related to your research content, may also give you some inspiration. On other occasions, you may even say to yourself: "This is so obvious (now that my colleague pointed it out), why didn't I think of it before?" The same thing might have occurred to you, in fact, when you read the reviewers' comments. In this context, it would also be very helpful if you have a chance to get away from the completed revision for a little while and come back to it and read the entire proposal as if it were someone else's proposal. Often you may be surprised that you find something new that needs reconsideration or reformulation. Incidentally, this "cooling-off" period is often necessary in the general context of creativity, when the problem solver must get out of the bias imposed by a fixed pattern due to continuous experience with a certain type of problem; cognitive psychologists have termed this type of creative problem solving "the Aha! moment," a moment when a creative solution jumps to mind instantaneously after a cooling-off period.

5.4 FINAL REMARKS ON REVISION

In revising a journal article for resubmission, the goal of the revision process is, regardless of whether you agree with the opinions of the reviewers, to convince the reviewers (and the editors) that (1) you took the comments and suggestions seriously and (2) you took steps to rethink the issues or re-run the experiments, and revise the exposition of ideas. For the revision of an NSF research proposal you should have much of the same goals, except that you may be dealing with more reviewers, and dealing with the Program Officer in a larger and more complex organization. The strategies discussed above suggest that you keep the reviewers' and Program Officer's perspectives in mind while revising your proposal, and aim at presenting a balanced, clear, and well-structured proposal with substantial reworking of the original proposal.

Although there is no guarantee that your revised submission will be successful for funding, we can draw the general conclusion that you are one step closer than you were last time to the funding line. It is possible that your revised proposal will be ranked worse than the previous version of the proposal, but this is unlikely if you have followed the above suggested action plans. What the investigator cannot determine, nor can the Program Officer, is that new reviewers or new panelists may be involved in reviewing your proposal, as a result of either panel turnover or the unavailability of previous reviewers who read your earlier proposal (see Chapter 4 on selection of panelists and reviewers). They may have a new perspective on your project different from that of previous reviewers, and if the new perspective is unfavorable to the general approach, you may indeed end up with a worse ranking the second time around. In such a case, you may be frustrated, but do not give up. Each time you revise and resubmit, you have uncertain factors that will swing the reviews one way or another, but in the end, each time you do the revision, you are in general closer to success, unless you get direct comments on the total inappropriateness or infeasibility of your research approach or methodology (see Sections 5.2.1 and 5.2.2). Unlike NIH, NSF does not prohibit you from submitting a proposal a third or a fourth time, so take a deep breath when you get bad reviews and keep revising and make your proposal better! This writing and rewriting process hopefully also makes you think more clearly about your research perspective and allows you to situate your work in a larger and broader social and scientific context.

A final word about contacting NSF for revision: It is perfectly fine to ask the Program Officer questions at any stage of the grant process, as discussed in Chapter 1, Section 1.3.1. This is especially helpful when you get encouraging reviews (but not final good news) and are about to work on the revision. Check with the Program Officer to see where your major efforts should go (see Sections 5.2.1 and 5.2.2). Program Officers get all kinds of inquiries, requests, and responses from investigators, and the least liked ones include messages from upset PIs who receive bad news and complain about the unfairness of the reviews. From the Program Officer's perspective, there is not a lot that can be done after a declination has been issued to the PI. In principle, the PI can file a formal complaint or appeal to NSF, arguing that serious deficit exists with the review process. Indeed, NSF does have a formal mechanism that allows investigators to request that their proposal be reconsidered (see the procedure listed in Chapter IV of the Grant Proposal Guide, Section D: Reconsideration; http://www.nsf.gov/pubs/policydocs/pappguide/nsf13001/gpg_4.jsp). In practice, arguments of this type are often difficult to verify and sustain, especially when examined within the context of budget availability and program priorities. Comments from a given ad hoc review represent only the individual researcher's perspective, and comments from a panel represent a collection of experts' views on the issue; unless there are serious wrongdoings in the review process (on the part of the reviewers or the Program Officer), it is unlikely that such formal complaints will lead to a positive result (see Overview of Section D, Chapter IV of the Grant Proposal Guide). In the meantime, the investigator may have burned some personal bridges between the investigator and the NSF Program Offier or the reviewers and panelists. In short, the best approach to getting your proposal funded is: revision, revision, and revision.

Chapter 6

Managing Your Grant

Seasoned grant awardees have a saying: "The good news is that your grant got funded. The bad news is that your grant got funded." The bad news referred to in this saying is that you are now responsible for managing your grant, producing the work promised, and situating yourself for the next project—not exactly bad news, but certainly a good deal of work (and a good problem to have!). Your first grant from NSF, or from any funding agency, will set the stage for your relationship with that funding agency. If you produce good work, publish and disseminate your results, and provide educational experiences for students, you will begin a long-term relationship with NSF. So, it is very important that you know how to manage your grant so that the important intellectual work can get done and get disseminated.

6.1 THE GOOD NEWS LETTER: NEGOTIATIONS WITH YOUR PROGRAM OFFICER

The process of making an award at NSF was discussed in Chapter 4. Part of a Program Officer's decision in making a recommendation might include follow-up negotiation with the PI. Negotiation may include further elaboration and clarification about parts of the proposal that were unclear, and often include negotiation about your proposed budget. Typically, the negotiations around budget are in budget reductions. Program Officers may ask you to reduce your budget by a particular percent (e.g., 10%–20%), further justify your proposed budget, align your budget with your proposed scope of work, or cut out some of your proposed work and as a result reduce your budget. A possible sample letter from an NSF Program Officer is shown below. Each program handles initial communication of the "good news" with PIs in different ways. The following letter is a sample that we created based on good news that the authors previously received as investigators, to give you a sense of what you might expect in terms of communication from an NSF Program Officer to investigators.

Having Success with NSF: A Practical Guide, First Edition. Ping Li and Karen Marrongelle.
© 2013 Wiley-Blackwell. Published 2013 by John Wiley & Sons, Inc.

Dear Dr. Lucky,

Your NSF proposal did well in our review process. Both reviewers and NSF staff who read your proposal felt that it was well written and that you have assembled a strong team of researchers and educators who are likely to successfully implement your project. Hence, we plan to recommend it for funding. I must point out that this is just a recommendation; only an NSF Grants Officer can make an award.

Some questions were raised by reviewers and Program Officers. I am hoping that you can return a set of responses to me in about a week and a half. Please let me know if you will not be able to respond within this time frame.

A. RESOURCES. Have there been any changes that affect the proposed project? (For example, change of affiliation, illness.) Has your access to resources changed? Is this proposal under review anywhere else, and are there any other funding resources covering any aspects of the proposed project? Has the project been funded elsewhere? Potential overlap in funding is a serious issue. If the answer to any of these questions is yes, please let me know, so that we can discuss the implications.

B. ABSTRACT. Prepare a draft abstract for the general public that describes the work to be undertaken and its significance and broader impacts. The abstract should consist of two short paragraphs aimed at getting someone like a congressperson (or your grandmother) excited about your work. The first paragraph should be a descriptive summary of the research project, and the second paragraph a non-technical explanation of the project's broader significance and importance, and how it might lead to fundamental scientific discoveries. No jargon please. The language should be as vivid and accessible as possible. Do not use special formatting such as italics or boldface. Do not use first person (use "the investigator" or "investigators" instead of "I" or "we"). The attached document provides more details about preparing the public abstract. You can find sample abstracts on the NSF website. Because abstracts are available to such a wide audience, high standards of quality must be maintained in preparing them.

C. HUMAN SUBJECTS. An up-to-date indication from your IRB (human subjects) that your research has been evaluated and approved needs to be emailed to me as a pdf file or a clear jpeg image. I can't recommend an award without it. In most cases, IRB approvals are good for one year. Your period of approval should be good at least for three months beyond your start date. If your approval will expire before that, you should get new or renewed approval. The title of the project indicated on the IRB approval letter must be the same as the title of your NSF proposal. If it is different, then have your IRB officer or Sponsored Research Office give you an official note explaining that the IRB approval refers to the same project (again, a pdf or clear jpeg is best, with the relevant official's signature). The IRB approval and any amendments must be on official university letterhead with the responsible official's signature. An electronic signature is not acceptable. Your IRB approvals need to be in before I can even recommend your project for funding.

D. GRANT AMOUNT: Given our tight budget, it is our program policy to use the available money to fund as many projects as possible, so we will not be able to fund your project at the level of your original request. In addition, reviewers raised questions about the size of your budget in relationship to the proposed research. I agree with them. Please prepare new budget pages and budget justification for a budget with a 20% reduction over the [x] years of the project. If international travel is requested, then the specific meeting and costs must be explicitly specified. In your revised budget justification, you will need to include a Budget Impact statement that explicitly details how the reduction in an award would impact the project. You will need to submit the revisions via FastLane, and your SRO must approve the submission.

E. REBUTTALS: I attached informal copies of your reviews and the Panel Summary. The reviews include discussion of concerns/issues that were raised in the course of the review. I would like you to prepare a rebuttal to the concerns/issues in the Panel Summary. The rebuttal will be uploaded to FastLane as permanent record associated with this project.

Sincerely,
Your NSF Program Officer

We will point out a few things about this sample "good news" letter. First, the sample letter asks the PI if there have been any changes that would affect the proposed project. Remember, by this point, it has been about six months since the proposal has been submitted. So what could change in six months? One possibility is that you had luck getting the same project funded by another agency at a higher amount! So you may want to tell NSF that you are withdrawing the proposal from them (which may make the Program Officer happier, because he or she can then use the money to support another worthy project!). Another possibility is that you may have accepted a job outside of academia and can no longer serve as the PI on the project. It may also be the case that nothing major has changed that would affect the project.

Second, the sample letter also asks for an abstract. All projects funded by NSF are required to submit an abstract for the general public. The abstract must address intellectual merit and broader impacts, but in language that can be understood by the layperson. Your project cannot be funded until an abstract is received and approved by your Program Officer. As a result, all "good news" letters from Program Officers will probably contain a request for an abstract or the Program Officer will ask you for an abstract at some point prior to the funding of your project. All abstracts are public; you can search for and view abstracts of NSF awards at the Awards Search Page on the NSF website at http://www.nsf.gov/awardsearch/. There are a number of ways to search NSF awards: one way is by a keyword or phrase that exemplifies your project and another way is by PI name. You can also further refine awards searches by searching within a particular division or directorate and/or by searching by your Program Officer's name. However you choose to search the Awards Search page, the results will provide information about the types of projects that a program has been funding and will provide example abstracts for your program. Exemplary abstracts include information about the intellectual merit and broader impacts, but digestible by the general public. As you start to disseminate the results of your project, your NSF Abstract Page will be updated with publications produced as a result of your research.

We also point out that the letter asks about human subjects research approval. If your project involves human or vertebrate animal subjects in your research, hopefully you checked the appropriate box on the cover sheet (see Chapter 3, Section 3.2.2.2) when you submitted your proposal (see also Chapter 2, Section 2.12, for discussion of human and vertebrate animal subjects). In some cases, you may have submitted an application to your Institutional Review Board (IRB) when you submitted or even before you submitted your proposal to NSF. In other cases, you may wait to submit your application to your IRB until you hear from NSF about the status of your proposal (note that this will delay the award time because IRB approval usually takes time, from a few days to a few weeks depending on the institution). No matter the status of your IRB application, when you receive word from NSF that your proposal is being recommended for funding, your Program Officer will ask you to send a copy of your IRB approval or exemption. You project cannot be funded unless NSF has a copy of an IRB approval or exemption letter, showing that your project was reviewed and either approved or deemed exempt from review.

The sample letter also asks the PI to respond to questions raised by the reviewers or panelists. In the sample letter, the Program Officer asks the PI to respond to issues raised in the Panel Summary. However, different Program Officers might approach asking you to respond to questions and issues raised by reviewers in a different way. For instance, some Program Officers may craft specific questions for the PI, based on the reviewers' comments. The Program Officer will provide a time frame, usually quite short, in which you must respond to the issues and questions raised by the reviewers. The Program Officer will review your responses and might ask you for further clarification or description. Although uncommon, a Program Officer might decide not to fund your proposal if your responses are not satisfactory. Hence, it is important to take seriously your responses to the questions raised by the review panel and Program Officer. In a few cases, your proposal may have received such enthusiastic and glowing comments that the Program Officer will need no further responses from you at this stage.

Finally, the letter also asks the PI to reduce the budget of the project. If you are asked to revise your budget (and most awards are made with a revised, not the original, budget), you want to ensure that the changes you make will not compromise the work that you promise to do. For instance, if you cut out a graduate student to reduce your budget, you want to think about the impacts of that loss on your ability to carry out the work of the project. Will other project staff be able to pick up the responsibilities envisioned for that graduate student? Will this delay the original timeline of the project? Reductions in any budget should be carefully considered: if you keep the graduate student and instead cut travel, how will the loss of travel funds impact your dissemination plan? In the spirit of seeking opportunities, when you face a reduction in your budget, you may need to get creative in how you fund the full scope of your planned work. For instance, you may be able to negotiate with your university to help pick up some of the costs that you will need to cut. For instance, your university might be able to furnish graduate student tuition waivers so that you can keep the graduate student on your project by only paying the graduate student stipend. Another place to look to cut may be your salary supplements.

Although this is disheartening, many investigators find that the only thing they can cut (without impacting the scope of the proposed work) is their own summer salary (e.g., reducing by 50%). Again, here you can also negotiate with your institution, to see if there are other revenues that the deans and chairs may tap into so that you can get some summer support. In short, when faced with having to cut your budget, think creatively and think flexibly!

Program Officers enter into negotiations with you and that's exactly what they are: negotiations! If you are asked to cut your budget by 10%, but can only cut by 7% without compromising your proposed work, talk to your Program Officer! Sometimes you can find places to cut in your budget, but other times making the requested cuts will result in compromising the work that you have planned. You should never promise to do work that is unrealistic for you to carry out with the resources you have requested (but also recall that you should not pad your budget when you first make the proposal; see Chapter 1, Section 1.3.4).

6.2 ENTERING BUDGET REVISIONS INTO FASTLANE

It is a good idea to revise your budget with your SRO. You or your SRO will then upload your revised budget into FastLane, using the "Revise Submitted Proposal Budget" tool under Proposal Functions in FastLane (see Figure 3.3 in Chapter 3). Any time you upload a revised budget, you will need to include a budget impact statement. The budget impact statement outlines how the changes to your budget will affect the work that you have proposed. Your SRO will need to submit the revised budget, just as your SRO submitted your completed proposal. A sample budget impact statement template is below.

Sample Budget Impact Statement Template

In response to NSF's suggestion we have derived a revised budget, in which the following major adjustments are made to the original budget: [list the changes to the original budget: (1)____, (2) ____, (3) ____, etc.].

The above revisions will not impact the overall scope and the intellectual content of the proposed research. [Insert a description of your reasons why the overall scope of the research will not be impacted. You might include arguments such as writing up results for publication can be done without substantial financial support, the PI taking reduced financial support without a reduction in effort, the PI's institution providing funds to cover graduate students' tuition, etc.]

6.3 MOVING ALONG THE RECOMMENDATION CHAIN

Your Program Officer will make a recommendation to fund your proposal at the negotiated budget rate. Once the Program Officer's recommendation is made, the Division Director must concur with the Program Officer. It is possible that the

Division Director will have additional questions for you to answer. Once the Division Director concurs with the Program Officer's recommendation, your proposal is forwarded to the Division of Grants and Agreements (DGA) for the awarding of your grant. Your institution will be notified by NSF's DGA once your proposal is funded. Be aware that the time between when your Program Officer recommends your project for funding and when the DGA makes the award can be weeks or longer, especially during the summer months. Then, it's time to get to work!

6.4 MANAGING YOUR GRANT

When you first hear the good news from your Program Officer, you should begin setting the stage for starting your project. This might mean hiring graduate students, hiring a project director (or lab manager), or talking to your department chairperson about course buy-outs. It also might mean reserving laboratory space or equipment time. If necessary, you are able to spend funds prior to the actual awarding of your grant; however, your institution assumes all risks associated with pre-award spending. That is to say that NSF is not obligated to make an award once it is recommended. You will want to consider the workload of the newly funded project in the context of your other work: teaching, supervision of students, other funded projects, committee service, and the like.

In many cases, with the exception of very large projects, you may be both administering the grant (that is, managing the project on a day-to-day basis) and conducting the intellectual work of the project. Both of these functions take time to do well. Depending on your institution, you may have quite a bit of support administering the grant, even without any support (i.e., do it yourself!). If you don't have administrative support, you will need to negotiate with your department chair or your dean to get other resources or to get release from other work you are doing (e.g., teaching and committee work).

6.5 BUDGETING AND THE MANY QUESTIONS ABOUT SPENDING MONEY

You should think of your budget as a best guess at how you plan to spend your funds over the life of your project. It is typical that things happen that leave deficits in some parts of your budget and surpluses in other parts. For instance, your salary or fringe benefit rate may increase between the time you submit your grant proposal and the time it is funded. You cannot request to increase the overall amount of the award to cover the difference in cost, but you can re-budget within the grant so that you get paid at your increased rate (but with caveats; see below). Believe it or not, it is usually the case that spending all of your grant dollars in the originally proposed amount of time for your project is challenging. This is one of the reasons why so many PIs request no-cost extensions: so that they can spend out their grant funds!

As we mentioned in Chapter 2 (Section 2.4.8), NSF does not have the manpower or resources to oversee the budget operation of most regular research projects once the funds are released to the institution. Thus, it leaves budget issues to the grantee institutions, and it is your and your institution's responsibility to monitor the budget and carry out the research in good faith. The General Grant Conditions have specific rules regarding when prior approval is required for change of budget or change of research scope. If you have a change of plan in using your funds (same money, different uses), you need to discuss this with your SRO staff and determine whether you will need to get NSF approval for it.

In general, you will not need NSF's approval for re-budgeting unless there is a change in objective or scope of your proposed research (see NSF's Grant General Conditions, http://www.nsf.gov/awards/managing/general_conditions.jsp; Article 2, Item b and Article 7). The two predominant conditions for which you will need NSF approval prior to re-budgeting are (1) moving funds from Participant Support Costs and (2) adding a subaward to your existing project (i.e., adding a subaward that was not outlined in the original grant proposal). If you have funds leftover in Participant Support Costs, you should discuss this with your Program Officer, as it usually signals some change to the project (see also Chapter 2, Section 2.4.8, item F, for details on Participant Support Costs). For instance, you are evaluating a professional development program for teachers and anticipate enrolling 90 teachers over the life of the project. However, by the last year of the project you enroll a total of 80, not 90 teachers. You then have funds left over in your Participant Support Costs category. You might propose to your Program Officer to use the leftover funds in the Participant Support Costs category to help fund a graduate student to disseminate project results at a conference. Your Program Officer will review with you the implications for enrolling fewer than the projected amount of teachers and together you will decide if and how to reallocate the Participant Support Costs funds. If you and your Program Officer decide that it is appropriate to reallocate your Participant Support Costs funds, you will enter a request to reallocate the funds through Fastlane's Notifications and Requests module (which is under "Award and Reporting Functions"; see Figure 3.2 in Chapter 3). You will need to specify that you are requesting to move funds from the Participant Support Costs category and you will need to indicate which budget category you want to move the funds to. You will also need to explain the reason for the reallocation.

A similar process should be followed for adding a subaward and reallocating the budget to accommodate the subaward. We discuss situations in which a subaward might be added to a project after your project has been awarded in the next section.

6.6 CHANGES TO THE PROJECT

Small changes in your work plan do not need to be approved by your Program Officer, but there are some major changes that require approval from NSF. If there is a significant change in the scope of work of your project, you should talk to your Program Officer immediately. Some significant changes are discussed above in Section 6.5 and in the remainder of this section; however, if you are unsure whether a proposed change is or is not significant, ask your Program Officer. Program Officers realize that research projects never go exactly as planned and so changes, mostly small, but sometimes large, happen along the way. If you and your Program Officer determine that a significant change to the scope of work is necessary and makes the best sense for completing the work proposed, you will need to obtain NSF approval for the change. You will need to use the NSF Notifications and Requests feature in FastLane to submit a statement of the change in the scope of work. As the PI, you will typically prepare the notification and your SRO will sign off on it.

One large change is the addition of a subaward. If you decide to add a subaward to your project, that is, contract with another organization to conduct some significant part of the work (one that was not named in the original proposal), you will need to get prior approval from NSF. For instance, you promise to create assessment instruments as part of your materials development project, but quickly become overwhelmed with the assessment development process. You meet an assessment expert at a conference in your field and determine that she would complete an assessment portion of your work better than you can. In order to contract with her, through her university, to complete the assessment development for your project, you must re-budget. Because you will be creating a subaward for the assessment development work, NSF must approve the change in scope of work (in this case, who will be conducting the work). Any time that you request a subaward be added to the project, you must provide the following information to NSF: (1) the reason for the subaward, (2) the process by which the proposed subawardee was chosen, and (3) a budget for the subawardee. Note that the subaward must come from your original budget funded by NSF; that is, NSF will not give you new money for this.

Another large change involves the Principle Investigator's relationship to the project. Typically, changes in the PI's relationship to the project fall into one of three categories: (1) devoting less effort than originally anticipated to the project, (2) changing institutions, or (3) short or long term absence of the PI. We will discuss each of these scenarios in order.

Seasoned PIs submit a number of grant proposals around the same time to NSF and other funding agencies (see discussion in Chapter 7, Section 7.1). It is not impossible that you have more than one proposal to get funded. When multiple proposals get funded at the same time, the original effort you anticipated for one or more of the projects might need to be reduced. If you need to reduce the amount of time allocated to a project by 25% or more, you will need to get your Program Officer's approval. This change, like other substantive changes, will need to be requested through FastLane's Notifications and Requests module. Your SRO will need to sign off on the request and your Program Office will need to approve it.

It is not uncommon that PIs change institutions during the life of a project and sometimes even before a grant is awarded. Because a grant is awarded to an institution, not an individual, the institution from which the PI is leaving must initiate a transfer of the grant to the PI's new institution. Typically, this transfer process is amicable, because institutions do not want to try to recruit a PI for an existing project. If you are changing institutions, you will need to contact the SRO at your current institution and the SRO at the institution to which you are moving. If both institutions and NSF are agreeable to you moving the project to your new institution, you should initiate a new Notifications and Requests function via the NSF FastLane system by clicking on the PI transfer button under Grantee Request Types. Your request should outline: (1) the work done on the project to date, (2) a description of the work that will be accomplished, (3) an estimate of the amount of funds left in the grant, and (4) a detailed budget for the amount of the grant you want to transfer to your new institution (typically, you will want to transfer the balance of your grant to your new institution). Your original institution will need to submit a report verifying the amount of funds left in your grant and to sign off on the request. Additionally, the SRO at your new institution will need to submit a detailed budget that matches the budget you submitted and will also need to sign off on the request.

Keep in mind that neither institution is required to go along with the transfer, so it will be incumbent on you as the PI to negotiate with the institution you are leaving and the institution to which you are going to take your funded projects with you. Occasionally, the transfer may be complicated by the type of grant you receive, or the type of institutions involved. For example, if you received an NSF grant while you were at a primarily undergraduate institution (e.g., a liberal arts college) and you are now moving to a research university (e.g., doctorate-granting university with very high research activity according to the Carnegie classification; see http://classifications.carnegiefoundation.org/descriptions/basic.php), you need to consult NSF and get permission to make the transfer. This is because your funded project may be under the category RUI (Research for Undergraduate Institution), and it has a significant component for training undergraduate students in conducting research. Moving the grant to a research-intensive university may not fulfill some of the original goals of the project. In such cases, one solution might be to find a PI at the institution you are leaving to act on your behalf to train the undergraduate students as originally proposed, and ask the institution to make a subaward to the new institution you are moving to, for just the part of the research that can be conducted at the new institution. Of course, if you are moving from one undergraduate institution to another institution of a similar type, there would be no problem for the transfer, as is usually the case.

Finally, if you as the PI must take time away from the project for less than three months (e.g., for a maternity leave or a sick leave), you should notify the Program Officer and discuss arrangements for the continuation of the project in your absence. If you must take time away from the project for more than three months (e.g., to serve on a rotation at NSF as a Program Officer) and intends to return, then you must provide written notification to the Program Officer about how the project will continue in your absence. Typically, this involves naming another PI or co-PI to the project to conduct the work while you are away. The "substitute PI" may need to work closely with your students or postdoctoral fellows so that the original work can be carried out as planned.

6.7 SUBMITTING ANNUAL, FINAL, AND PROJECT OUTCOMES REPORT

NSF requires project PIs to submit annual reports, regardless of whether your award is a *standard grant* (all of the funds for all years of the project are released in Year 1) or a *continuing grant* (the funds for Year 1 only are released in Year 1, the funds for Year 2 are released upon the Program Officer's assessment of satisfactory progress in Year 1, and so on for the subsequent years of the grant). You will receive an automatic reminder to submit your annual report 90 days before the anniversary date of your grant award. Annual reports should be submitted 90 days before your grant's anniversary date comes up. The final report, however, is not due until 90 days after the end of your project; the end date is specified on the cover sheet of your original proposal (or any other new date approved by the Program Officer due to No-Cost Extension; see the next section). It is important to submit your annual and final reports on time, especially if you have a continuing grant or another grant under consideration by NSF. A Program Officer cannot release your subsequent year's funds without reviewing and approving your annual report for the current year. It is equally important to submit your annual or final report on time if you have a standard grant, as late reports can hold up funding for you or your co-PI's other, pending grants. As a Program Officer, it would be frustrating to try to make a recommendation for funding and have the action held up by a late annual or final report from a co-PI. Please note that if the Program Officer finds your annual or final report unsatisfactory, he or she will typically first ask you questions, check with you for clarifications, or request a revised report. If the Program Officer remains unsatisfied, he or she can decline further funding or request your institution to return the remaining funds to the U.S. Treasury (the funds do not go back to NSF!). This latter situation occurs whenever NSF detects a fraud associated with the use of NSF funds or a misconduct of research associated with the project.

The annual report is one important way to communicate with your Program Officer about the progress of your project and any changes that you anticipate making to your work plan. Of course, it is helpful to communicate with your Program Officer other times throughout the year about any questions, problems, or highlights that arise as you carry out your work. Your annual and final reports will be submitted in FastLane and will follow a prescribed format that asks for items such as the number of research papers published, talks given at conferences, person-time devoted to the project by project personnel. The sections of the Annual Report are as follows:

1. Project Participants: In this section, you list by name the senior personnel, postdoctoral researchers, graduate students, undergraduate students, technicians, and other participants who contributed 160 hours or more during the year to the project.
2. Organizational Partners: In this section, you list other institutions that participated in the project (typically your subawardees) with a description of the people involved and work completed.
3. Other Collaborators or Contacts: In this section, you describe other participants in the project who were not working under a subaward agreement.
4. Activities and Findings: In this section, you describe the major research and educational activities, deliverables produced, results garnered, and Broader Impacts activities. NSF now requires that annual and final reports address progress in all activities of the project, including any activities intended to address the Broader Impacts

criterion that are not intrinsic to the research. Thus, you need to describe here how your activities achieve the proposed goals of intellectual merit and broader impacts.

5. Journal Publications: In this section, you list the pending and published peer-reviewed manuscripts based on the research and educational activities of the project during the previous year. You will do the same for books and internet sites (e.g., web-based tools for research or dissemination).
6. Contributions: In this section, you will describe the contributions of the project to your discipline, to other disciplines, to human resource development, to research and education, and to society.
7. Conference Proceedings: In this section, you will list any conference proceedings associated with the project.

You may include appendices in your annual report to demonstrate the type of product produced or a particular article of interest, typically as attached files in PDF format. Your final report should describe the culmination of your project's activities in the final year of the project and will follow the same reporting categories as described in the annual report above.

NSF now requires a Project Outcomes Report to be submitted via Research.gov. The Project Outcomes Report is a brief summary (200–800 words) written for the general public that provides an overview of the major results and findings of the project (think about this as writing the abstract of your project, but for your completed project). Like the annual and final reports, the Project Outcomes Report should describe both the intellectual merit and broader impacts of the project. The Project Outcomes Report is submitted in addition to the final report. Like the annual and final reports, failure to submit the Project Outcomes Report in a timely manner may result in a delay in review or processing of proposals for you or your co-PIs.

6.8 NO-COST EXTENSION

Believe it or not, it is often difficult to spend all of the funds in your grant within the period of time that your project is funded. This happens for a number of reasons: for example, you are careful when spending grant funds, so as not to spend over budget, or you might start your project some months after the "official" word comes from NSF due to timing or taking longer to line up people to work on the project. If the end of your project is approaching and you have funds remaining in your project budget with work yet to complete (analysis of data, writing papers for publication), you can notify NSF of a no-cost extension to finish your work. You will notice that this so-called "grantee approved" no-cost extension is *authorized*, not *requested*, by the PI's institution. This is because NSF allows the grantee institution to extend the grant period by up to one year if there are funds remaining and work to do (remember, funds that are returned go to the U.S. Treasury, not back to NSF). A notification of a grantee-approved no-cost extension must be filed on FastLane (using the trusty Notifications and Requests function; see Section 6.5) at least 10 days prior to the expiration date of your project. However, the earlier you submit a no-cost extension authorization, the better. Your Program Officer will acknowledge that he or she received your notification of the no-cost extension. You should also be prepared to submit your annual report soon after you enter your notification of a no-cost extension. Until you submit your no-cost extension, FastLane thinks that you are in your final year of the project and so your final report would normally be due 90 days after your grant expiration date. However, once you trigger a no-cost extension and your Program Officer acknowledges your notification, the expiration date on your grant will change and you will have an annual report due.

In exceptional cases, you can request a second no-cost extension if you have funds remaining and work yet to be completed. You will again make a request in the Notifications and Requests section of FastLane; however, you will need to provide a rationale for the extension request and a plan to spend your remaining funds. This time the request must be received by NSF at least 45 days prior to the grant's expiration date. Your Program Officer will need to approve your request, so it is a good idea to talk with your Program Officer before submitting a request through FastLane.

You should check with your Program Officer to be sure that you will not incur negative repercussions as a result of extending your project via a no-cost extension. For instance, some programs that make awards in phases may not allow you to apply for a subsequent phase of funding while you are working on a project with a no-cost extension.

6.9 COMMUNICATION IS KEY

All research projects encounter bumps in the road, and you can better manage these bumps by communicating with your Program Officer. Your Program Officer can draw from her or his own experience in leading projects and handling the myriad of situations encountered when managing a portfolio of grant-funded projects, and thefore can provide you with advice and guidance. As we mentioned in Chapter 1, Program Officers are there to help you, and they are receptive to your inquiries and can in most cases respond to you promptly (see Chapter 1, Section 1.3.1). Your Program Officer has an interest in the outcome of your project, so keep your Program Officer informed!

Chapter 7

Extending the Horizon

7.1 WHAT'S NEXT

You have now successfully finished your first NSF grant proposal, received the funding, and executed part of your project. Hooray! Some of you may say you can rest a bit now. Unfortunately, you can't! In today's research environment, most likely you cannot afford to ride on the pride of one funded project and take a break. You must continue to think ahead for the next project while you are working on the current one. This is unlike writing articles, where you can say you want to work on one article at a time. For the *sustainability* of your research and the *continuity* of research funding, you should consider (1) how to further expand the current project so that it can leverage other resources and funding to support additional components of the project, (2) how to develop new lines of thinking on the basis of the current project so that you can receive future funding, and (3) how to develop new lines of research independently of the current project for future funding. This last point is also very important, because funding agencies often want to see innovative projects (NSF's *transformative concepts* review element; see Chapter 2, Section 2.5.1) and new groundbreaking (paradigm-shifting) ideas, rather than simple continuation or expansion of existing research. This emphasis on novel and transformative work is partly due to pressure from the government and private sectors, especially when there is not enough funding for many good projects given the budget situation.

In short, you must keep a pipeline of research ideas and projects, and keep writing grant proposals. Many investigators simply cannot afford a break in grant funding if they want keep their research programs running smoothly (and keep the students and researchers in the lab). Considering the long cycle of review process (see Chapter 1, Section 1.3.5) and the low rate of funding, it is all the more important that you always have grant proposals in the making or under review.

In this final chapter we discuss ways in which you can use your currently funded project to leverage additional resources, to develop new lines of research, and to expand the scope of your work in general.

Having Success with NSF: A Practical Guide, First Edition. Ping Li and Karen Marrongelle.
© 2013 Wiley-Blackwell. Published 2013 by John Wiley & Sons, Inc.

7.2 LEVERAGE FURTHER RESOURCES

7.2.1 Within NSF

There are mechanisms within NSF that can provide additional support to your currently funded project. Depending on the research objectives of your study, you might need to have additional support to personnel or to infrastructure. First, with regard to personnel, for example, you may want to involve motivated and dedicated undergraduate students to work on the project (in addition to graduate students already on your budget). NSF has a Research Experience for Undergraduates (REU) program, which provides supplemental funding to existing projects through the REU Supplements mechanism. REU Supplements must be requested by researchers of ongoing NSF-funded research projects, and the primary criterion for such funding would be the professional and educational benefits to undergraduate students in conducting research with you (see Chapter 2, Section 2.9, for details). The student's experience and qualifications, as well as his or her fit to project needs, will also be considered by NSF. Such requests must be made to the NSF program that funded your original project, and are normally assessed without external review. A typical type of funding for REU Supplements would be summer stipends, for example, in the range of a few thousand dollars for each student involved, in support of salary, housing, and other fringe benefits. A small amount of materials and supplies associated with the student's expenses may also be requested, although such expenses should be kept at a minimum.

Other than the involvement of undergraduates, you may also need additional graduate students to be involved in your project, and NSF has several mechanisms to support graduate students, apart from the usual support that one can apply for within a regular grant proposal. In Chapter 1, Section 1.3.3, we mentioned the GRF (Graduate Research Fellowship), which is provided to entry-level (first- and second-year) graduate students. These fellowships must be applied for independently by the students who are working or planning to work with you. There are also the Doctoral Dissertation Research Improvement grants that students at an advanced level of graduate study (ready to complete a dissertation) can apply for. Unlike the GRFs, which have to be submitted by the graduate student, the Doctoral Dissertation Research Improvement proposal must be submitted by the faculty member serving as the PI and the student as the co-PI. In this case, it is like a full-blown research proposal, with more details about the theory and experimentation of the dissertation study, in contrast to the two-page research proposal for a GRF.

If you find that your project has grown to a level where you feel you need a postdoctoral fellow, you might not realize that the program that funded your research actually has specialized postdoctoral fellowships you can apply for (see Chapter 1, Section 1.3.3). Again, these are opportunities that can be independent of the actual funded project, but it makes a stronger case if you can link a potential postdoctoral research to the current research funded by NSF. In some cases, there are specialized opportunities, such as the Postdoctoral Research Fellowship (SPRF) in the Directorate for Social, Behavioral, and Economic Sciences, which is designed to broaden the participation of scientists from under-represented groups and to encourage interdisciplinary research (see NSF 12-591 for details; see also Chapter 1, Section 1.3.3.3). Such fellowships need to be arranged before submission between you as a sponsoring scientist and the potential postdoctoral fellow, and could be submitted either by you or by the postdoctoral applicant as the PI, depending on the specific solicitations. It is important to keep in mind that a good match between the applicant's experience and your research is essential for the success of such postdoctoral fellowships.

In addition to the personnel needs that you did not foresee at the time of the original proposal, your project may also have grown to such a level that you need to have additional equipment. NSF does not have a separate mechanism for equipment supplements (like the REU supplements), so requesting additional equipment for the existing project normally does not work. However, if your equipment needs are on a large scale, particularly if they require tens or hundreds of thousands of dollars to purchase, you should consider writing a separate proposal to apply for funding from the Major Research Instrumentation (MRI) program (see http://www.nsf.gov/od/oia/programs/mri/). This is a case when your project must expand significantly to include new ideas, new lines of research, or new collaborations that make the inclusion of new equipment necessary. In other cases, such expansion of research ideas may lead to entirely new projects (see Section 7.3, below).

Finally, other than personnel and equipment, your research may have expanded to a level at which you feel that a special workshop with invited experts may provide an integrative view of the issues at hand. In this case, you may want to propose to hold a specialized conference or workshop in connection with your research project (see Chapter 1, Section 1.3.3). Note that your conference must provide a venue that is not typically available to the participants—in other words, the invited participants typically may not have an opportunity to meet in regularly held conferences organized by the disciplinary societies. In Chapter 1 we also mentioned several other types of support, some of which are one-time or emergent, that you can seek to further expand your work in research, training, and education. It is important to periodically scan the NSF website (or carefully read the NSF updates that come to your inbox; see Chapter 1, Section 1.2.3, for subscribing to NSF alerts) as NSF constantly rolls out new initiatives and programs, particularly in areas for cross-disciplinary and international collaborations, to which you should consider relating your current work.

7.2.2 Beyond NSF

Although it is beyond the scope of the current book to discuss what opportunities there are outside NSF, it is only natural that you connect your research to funding opportunities at other funding agencies. If your current work has practical implications for application in the military context, you may want to consider applications to the Department of Defense (e.g., through DARPA, http://www.darpa.mil/, the Defense Advanced Research Projects Agency; the ONR, http://www.onr.navy.mil/, the Office of Naval Research; or the AFOSR, http://www.wpafb.af.mil/afrl/afosr/, the Air Force Office of Scientific Research). If your work is related to education, you may be interested in expanding your work so that you can seek funding from the Institute of Education through the Institute of Education Sciences (http://ies.ed.gov/funding/). And, of course, if your work has health-related implications, such as in clinical diagnosis and intervention of disorders, you should expand it to projects that are suitable for consideration by the National Institutes of Health through its Office of Extramural Research (http://grants.nih.gov/grants/oer.htm). It is also important to note here that NSF often collaborates with these and other federal agencies to jointly support innovative research projects. See Chapter 1, Section 1.2.1, for discussion on NSF–NIH collaboration, and the website http://www.nsf.gov/about/partners/fedagencies.jsp for additional pointers.

There are also many other private funding agencies, such as the James McDonnell Foundation (http://www.jsmf.org), the Alfred Sloan Foundation (http://www.sloan.org/), and the Bill and Melinda Gates Foundation (http://www.gatesfoundation.org/Pages/home.aspx), to which your research may relate, and the key is for you to see what their priorities are and what aspects of your research can contribute to their missions and goals.

There may also be other international funding organizations, including the counterparts of NSF in other countries that you and your international collaborators may be interested in (e.g., the NSFC in China, http://www.nsfc.gov.cn/e_nsfc/desktop/zn/0101.htm; the European Science Foundation, http://www.esf.org/; the Korea Research Foundation, http://www.krf.or.kr/KHPapp/eng/maina.jsp; and the Human Frontier Science Program, http://www.hfsp.org/). Each of these funding agencies has its own priorities, criteria, and eligibility, and it is up to you to see how your research might fit in and appeal to the relevant agencies. A concrete step to take for international funding would be to collaborate with colleagues in a country where you plan to apply for funding, and NSF is highly supportive of such joint efforts. Indeed, as discussed in Chapter 1, NSF has mechanisms that specifically encourage such international collaborations through its Office of International Science and Education (e.g., the PIRE program, http://www.nsf.gov/funding/pgm_summ.jsp?pims_id=12819), and through newly created programs in various disciplines such as the NSF–DFG Collaborative Research (NSF 11-053) between United States and German scientists (see Chapter 1, Section 1.3.2) and the Open Research Area initiative supported by NSF and its counterparts in France, Germany, and the Netherlands (see Section 7.3.2 below).

7.3 FROM SMALL PROJECTS TO LARGE-SCALE STUDIES

7.3.1 "Rome Was Not Built in a Day"

Assuming that after various attempts at grant proposal writing, you have succeeded in receiving one or several grants from NSF or other federal funding agencies. In the most likely scenario, your first projects funded by NSF are at a smaller scale, limited to one or several individual studies or a group of well-connected studies. This is the right level to begin with, especially for investigators at their early career stages.

There are two important notes to make here. First, this does not mean that early-career investigators cannot develop large-scale interdisciplinary work; it simply means that it is natural for investigators to start with smaller-scale projects and then expand the research to large-scale studies. Strategically, this makes better sense if you are at an early stage of your academic career to write smaller grants (with modest research scope and modest budget) that you can use to bootstrap your work. Second, the authors of this book have noticed an interesting divide between NSF and NIH investigator-initiated research proposals. Most of the standard research projects funded by NSF, because of limited budgets, tend to be smaller than the NIH standard research projects (the R01s; NIH does have separate mechanisms to encourage smaller-scale projects, e.g., the R03s). It is often, if not always, the case that the NIH R01 research projects are demanded of a large theoretical framework that guides the research, and typically, early-career researchers are disadvantaged when their projects are evaluated alongside the projects of senior researchers against the theory criterion. The point here is that as an early-career researcher, you should let your initial projects naturally expand and grow to incorporate more ideas, more theoretical perspectives, and more convergent methodologies, so that in the end the research will build up to a level that has an overarching theoretical framework or a big and bold theoretical hypothesis. "Rome was not built in a day," and this saying applies very well to this context.

Once you have successfully executed a few small-scale projects, it is indeed time for you to think about large-scale studies. In Section 7.2 we have already mentioned ways in which your project may expand to include additional personnel and equipment so that you can leverage further resources. Here we want you to consider other ways to tap into opportunities within and beyond NSF. Within NSF, you can consider being a PI or co-PI on a large interdisciplinary research project, and depending on the area of research, NSF may have relevant million dollar-level collaborative projects you can apply for; for example, Cyber-Enabled Discovery and Innovation (CDI), Sustainable Digital Data Preservation and Access Network Partners (DataNet), Collaborative Research in Computational Neuroscience (CRCNS), and Science of Learning Centers (SLC) are all large-scale programs that are designed to enhance learning and discovery in cross-disciplinary and collaborative research environments. Some of these programs may be one-time, or temporary for a few years, due to emergent funds from various sources. Other programs may be more permanent, such as programs of Major Research Instrumentation (MRI), Science and Technology Centers (STC), Integrative Graduate Education and Research Traineeship (IGERT), and Partnership for International Research and Education (PIRE). These programs, MRI, STC, IGERT, and PIRE, may support projects at the multi-million dollar level, and require you as a PI to include substantial intercommunity, interdisciplinary, and international collaborations to bring research and education to fruition. Some of these projects, in keeping with NSF's mission of integrating learning, discovery, and innovation, may have strong emphases on training and education, such as the IGERT and PIRE, or on building research infrastructure or communities, such as MRI and STC (see also Chapter 1, Section 1.3.3; see also the next section for more discussion on PIRE).

7.3.2 Interdisciplinary and International Collaborations

To move from small- to large-scale projects, a key factor for success is to involve interdisciplinary and international collaborations. In Chapter 1 we mentioned how NSF places its priorities upon interdisciplinary studies and international collaborative research, given the complex science and technology challenges that we face globally. For most NSF programs, several researchers can submit one single proposal from multiple institutions, in the form of Collaborative Research projects (see Chapter 2, Section 2.7, and Chapter 5, Section 5.3.5). For some NSF programs, multi-site, multi-research team, collaborative work is a prerequisite (e.g., Cyber-Enabled Discovery and Innovation, Type II grant; NSF 11-502). Today's science has come to a stage where it is increasingly impossible for a single individual to carry out a study from beginning to end just by him- or herself. The complexity of modern scientific problems, and the theories and methodologies involved in dealing with these problems, often makes it a necessity to use teamwork, to collaborate, and to adopt each other's perspectives. Realizing these challenges, NSF continuously pushes out new initiatives and programs to promote interdisciplinary research, as has been discussed throughout this book (see another recent example on Interdisciplinary Behavioral and Social Science [IBSS] research competition; NSF 12-614).

In Chapter 5, Section 5.3.5, we provided some details about how to involve collaborators, and the attention needed to the synergy and expertise each collaborator brings to the table. Here we simply want to point out that when you move to the stage of attempting to answer big science questions, it is only natural that you seek out opportunities for large-scale research collaboration with investigators who have unique and complementary expertise in the area of your research. Of course, as you progress in your research career trajectory, you will naturally build relationships with other researchers and meet investigators across the spectrum of disciplinary fields. As such, developing large, multi-site, multi-research team collaborative work may be a natural step in your own career trajectory.

NSF's emphasis on international collaboration is reflected in the various new initiatives for enhancing joint work by U.S. scientists and scientists from other countries, including the various agreements signed by NSF and its counterparts in other countries, for example, the NSF–DFG Collaborative Research as mentioned earlier, the U.S.–China Collaborative Research in Advanced Sensors and Bio-inspired Technology, and U.S.–China Collaborative Mathematical Research, to mention just a few. Another recent example is the Dear Colleague Letter (NSF 13-005) demonstrating NSF's effort to reduce barriers to international research collaboration. The NSF's Directorate for Social, Behavioral and Economic Sciences is now participating in the Open Research Area (OPA), an initiative by NSF's counterparts in France, Germany, and the Netherlands to promote multinational collaboration in research. What OPA will do is to allow researchers from any combination of three countries to submit a single proposal to NSF or its counterpart, and go through a single review process such as the one at NSF for consideration of funding. The proposed project must demonstrate the added value of transnational collaboration, not simply having three researchers from three countries working on a project.

Perhaps the most well-known international collaboration program is the Partnership for International Research and Education (PIRE, NSF 11-564), which supports the collaboration between U.S. scientists and students to do joint work with international partners. With PIRE support, you can send students and researchers to spend an extended period of time in your international partner's institution and research labs, to collect data and run joint studies. This large-scale program (funded usually at the level of 3–5 million dollars) provides a mechanism for both scientific research and student training at multiple levels. An example of a PIRE project can be seen from Figure 5.1 in Chapter 5, which shows the collaborative and integrative research themes, cross-disciplinary and complimentary methodologies, and a large network of international partners.

In Section 7.2.2, we mentioned a few private foundations that may provide further opportunities for your research. Some of these foundations may also encourage strong international collaboration, such as the Human Frontier Science Program (HFSP; in fact, one of the authors of this book first came to the United States as a postdoctoral fellow with the support of the HFSP). A final, important point to note is that although the level of funding in the United States has been shrinking as the budget is tight in the past years, you may be surprised to see the level of resources that your international research partners have received, and by virtue of collaboration with your colleagues, you may be able to tap into other countries' resources that are normally not available to U.S. scientists.

7.4 GETTING YOURSELF INVOLVED

In this final section, we want to point out that it is important to get yourself involved with the granting process, including being a reviewer, being a panelist, or even being a Program Officer at the NSF or other funding agencies. The quality of the NSF review process relies on the active service of expert researchers in the field, and, as such, the quality of your research is the most important thing for getting funded at the NSF. At the same time, however, as we have seen in this book, it is also important to know how to communicate your research to the reviewers, panelists, and Program Officers. Sometimes unless you have been on "the other side," you may not know your own side well. Service to the NSF as a reviewer, a panelist, or a Program Officer allows you to see first-hand how the NSF review process works, what common problems the submitted proposals may have, what counts as the most fundable research, and what the latest research trends, methodologies, and perspectives are, as we have discussed in this book; finally, it allows you to meet other colleagues and experts in your field. Such experiences could be highly valuable to your own research, for example, in writing successful grant proposals in the future. Both authors of this book have been on the other side as Program Officers (and reviewers and panelists), and we feel strongly that you should get yourself involved, once you have successfully executed a funded research project.

Let's start with a negative example. Your research has grown tremendously, and you say that you are too busy to do anything else, so you turn down all the review requests, panelist invitations, and other opportunities of service to the NSF. What we must remind you is that by doing so, you are not only not paying the necessary dues to the field, but also sending a poor image of yourself to your colleagues and Program Officers at NSF. A real example (needing no details here) is that one NSF Program Officer was unhappy with an investigator who just received an NSF award but who turned down an ad hoc review request for another proposal. This NSF Program Officer wrote an email to the investigator, and said something to the following effect (not a verbatim repetition): "I'm very disappointed at your turning down my review request. Please note that a group of five expert reviewers dedicated their precious time to diligently review your own proposal that was just funded, not to mention the many hours that our panelists and other program staff at NSF have spent in handling your proposal from beginning to end. Please reconsider your decision." All of this is true, and the investigator was embarrassed and did agree to the review request in the end. Now, this does not mean that you have to accept a review invitation each time you are asked, and people will understand that there are circumstances under which it is simply impossible for you to do so, just as editors of journals understand that not every expert is able to review a manuscript every time he or she is asked to do so.

If you do decide to contribute to "the other side," how do you go about doing it? In many cases this may be automatic, since the Program Officers at NSF (or other funding agencies) may contact you directly, given your research expertise and your previous success at NSF (or other agencies). In other cases, you may volunteer to serve as a reviewer, a panelist, or a Program Officer (the last requiring a formal application). You could write to an NSF Program Officer, saying that you have experience and expertise in one or several areas of research, and you will be available as either an ad hoc reviewer or a panelist should they need you (see Chapter 4, Section 4.2, on the selection of ad hoc reviewers and panelists). You should also attach a short resume or a complete CV of yours in your email. Most Program Officers at NSF welcome volunteer gestures like this, and will seriously consider such requests.

NSF also holds a Regional Grants Conference twice a year, and the conference is mostly hosted by a regional or a national university in the United States (search for "regional grants conference" on http://www.nsf.gov/). Attendance to this conference is highly recommended, as many representatives from each NSF directorate and many programs will be there to give talks on new programs and initiatives, cross-disciplinary programs, issues on proposal preparation, the NSF merit review process, and future directions and funding priorities. Conference participants will receive a large amount of information about NSF and the funding process within a short period of time (2 days). In addition, participants can also get a chance to meet some Program Officers of their own disciplines, exchange ideas and information with other participants who may have previous NSF experience, and do a bit of networking and get to know the NSF staff and other colleagues personally. This would also be an occasion where you can directly ask NSF staff questions, some of which may not have answers on paper, including many issues that are discussed in this book. It will also give you a chance to find out how you may get involved in the review process as an ad hoc reviewer or panelist, should you present yourself as an expert in a field. Even if your own disciplinary Program Officer is not around, other NSF staff are very receptive to pass your information on to relevant colleagues (your own Program Officer may be only a few steps down the hallway from the person you met!).

We must include a final word about being an NSF Program Officer (see the Preface and Chapter 4, Section 4.2.3, for a description of Program Officer roles at NSF). This book grew out of the experiences of two investigators who have previously served as NSF Program Officers as well as PIs and co-PIs on NSF-funded projects. As we can tell you, if your time and experience allow, it is a very rewarding process to work for the NSF, as "rotators" who stay for a year or two and then rotate out (some eventually become permanent program managers and never rotate out). The rotator experience allows you to see how the entire NSF funding process works (as discussed in this book), and in addition, you are often required to participate in other cross-disciplinary programs, including putting forward and drafting new initiative proposals or Program Solicitations. You are frequently working with other colleagues within the directorate and across the foundation by serving on various committees and task forces or by coordinating joint panels. Occasionally, you will also interact with colleagues from other funding agencies, which may be down a few blocks from the NSF building (e.g., Office of Naval Research (ONR) and Air Force Office for Scientific Research (AFOSR)). These activities allow you to see how science is being conducted not only in your own field but also in other domains, not to mention the scientific benefits of reviewing research proposals by top-notch scholars even before the research ideas are executed (as opposed to seeing the output of research in papers or conference presentations some months or years later). In short, NSF Program Officer activities provide one with the opportunity to see the complexity and the big picture of science and education, as well as the operation of a major federal organization. As a result, one understands better the larger picture of science and science policy, becomes an expert in communicating science broadly, and becomes comfortable with working with many people of different perspectives. These abilities may in the future help you extend the horizon in leveraging further funding and resources and in expanding research ideas and interdisciplinary and international collaborations.

Index

Having Success with NSF: A Practical Guide, First Edition. Ping Li and Karen Marrongelle.
© 2013 Wiley-Blackwell. Published 2013 by John Wiley & Sons, Inc.